Life-Cycle Greenhouse Gas Emissions of Commercial Buildings

This book develops a model to evaluate and assess life-cycle greenhouse gas emissions based on typical Australian commercial building design options. It also draws comparisons between some of the many green building rating tools that have been developed worldwide to support sustainable development. These include: Leadership in Energy and Environmental Design (LEED) by the United States Green Building Council (USGBC), Building Research Establishment Environmental Assessment Method (BREEAM) by the Building Research Establishment, Comprehensive Assessment System for Building Environmental Efficiency (CASBEE) by the Japanese Sustainable Building Consortium, and Green Star Environmental Rating System by the Green Building Council of Australia.

Life-cycle assessment (LCA), life-cycle energy consumption, and life-cycle greenhouse gas emissions form the three pillars of life-cycle studies, which have been used to evaluate environmental impacts of building construction. Assessment of the life-cycle greenhouse gas emissions of buildings is one of the significant obstacles in evaluating green building performance. This book explains the methodology for achieving points for the categories associated with reduction of greenhouse gas emissions in the Australian Green Star rating system. The model for the assessment uses GaBi 8.7 platform along with Visual Basic in Microsoft Excel and shows the relationship between the building's energy consumption and greenhouse gas emissions released during the lifetime of the building. The data gathered in the book also illustrates that the green building design and specifications are becoming more popular and are being increasingly utilized in Australia. This book is important reading for anyone interested in sustainable construction, green design and buildings, and LCA tools.

Cuong N. N. Tran is Lecturer at Ho Chi Minh City University of Technology (HCMUT), Ho Chi Minh City, Vietnam. He is currently Editorial Secretary of *International Journal of Construction Management* and Topic Editor of *Sustainability*.

Vivian W. Y. Tam is Professor, Associate Dean (Research and HDR), Associate Dean (International), and Discipline Leader (Construction Management) at School of Built Environment, Western Sydney University, Australia. She is currently Editor-in-Chief of *International Journal of Construction Management* and Senior Editor of *Construction and Building Materials*.

Khoa N. Le is Associate Professor at School of Engineering at Western Sydney University, Australia. He is Editor of *Journal on Computer Networks and Communications*, Hindawi Publishing, and Editor for *IEEE Transactions on Vehicular Technology* and *IET Signal Processing*.

Spon Research

Publishes a stream of advanced books for built environment researchers and professionals from one of the world's leading publishers. The ISSN for the Spon Research programme is ISSN 1940–7653 and the ISSN for the Spon Research E-book programme is ISSN 1940–8005

Making Sense of Innovation in the Built Environment
Natalya Sergeeva

The Connectivity of Innovation in the Construction Industry
Edited by Malena Ingemansson Havenvid, Åse Linné, Lena E. Bygballe and Chris Harty

Contract Law in the Construction Industry Context
Carl J. Circo

Corruption in Infrastructure Procurement
Emmanuel Kingsford Owusu and Albert P. C. Chan

Improving the Performance of Construction Industries for Developing Countries
Programmes, Initiatives, Achievements and Challenges
Edited by Pantaleo D Rwelamila and Rashid Abdul Aziz

Work Stress Induced Chronic Diseases in Construction
Discoveries Using Data Analytics
Imriyas Kamardeen

Life-Cycle Greenhouse Gas Emissions of Commercial Buildings
An Analysis for Green-Building Implementation Using A Green Star Rating System
Cuong N. N. Tran, Vivian W. Y. Tam and Khoa N. Le

For more information about the series, please visit: www.routledge.com/Spon-Research/book-series/SE0851

Life-Cycle Greenhouse Gas Emissions of Commercial Buildings

An Analysis for Green-Building Implementation Using a Green Star Rating System

Cuong N. N. Tran, Vivian W. Y. Tam, and Khoa N. Le

Routledge
Taylor & Francis Group
LONDON AND NEW YORK

First published 2022
by Routledge
2 Park Square, Milton Park, Abingdon, Oxon OX14 4RN

and by Routledge
605 Third Avenue, New York, NY 10158

Routledge is an imprint of the Taylor & Francis Group, an informa business

British Library Cataloguing-in-Publication Data
A catalogue record for this book is available from the British Library

Library of Congress Cataloging-in-Publication Data
A catalog record for this book has been requested

ISBN: 978-0-367-64685-1 (hbk)
ISBN: 978-0-367-65175-6 (pbk)
ISBN: 978-1-003-12819-9 (ebk)

Typeset in Goudy
by Apex CoVantage, LLC

Contents

Tables

Figures

Equations

Acronyms and abbreviations

AHIA-	Australasian Health Infrastructure Alliance
AQUA-	Alta Qualidade Ambiental
ASHRAE-	American Society of Heating Refrigerating and Air-Conditioning Engineers
AusHFG-	Australasian Health Facility Guidelines
BASIX-	Building Sustainability Index
BCA-	Building Code of Australia
BEAM-	Building Environmental Assessment Method
BREEAM-	Building Research Establishment Environmental Assessment Methodology
CALGreen-	California Green building Standards
CASBEE-	Comprehensive Assessment System for Building Environmental Efficiency
DGNB-	Deutsche Gesellschaft fur Nachchaltiges Bauen
DCB-	Dichlorobenzene
EEWS-	Ecology, Energy saving, Waste and Health
GBC-	Green Building Council
GBCA-	Green Building Council Australia
GDP-	Gross Domestic Product
GHG-	Greenhouse gases
GWP-	Global Warming Potential
GWP100-	100-year Global Warming Potential
HK-BEAM-	Hong Kong Building Environmental Assessment Method
HQE-	High Environmental Quality
IBPSA-	International Building Performance Simulation Association
ICC-	International Code Council
ICE-	Inventory of Energy and Carbon
IGBC-	Indian Green Building Council
IgCC-	International Green Construction Code
ILCD-	International Reference Life-cycle Data System
ISO-	International Organization for Standardization
LCA-	Life-cycle Assessment
LCG-	Life-cycle greenhouse gas emissions

LCI-	Life-cycle inventory
LCIA-	Life-cycle impact assessment
LEED-	Leadership in Energy and Environmental Design
MDGs-	Millennium Development Goals
NABERS-	National Australian Built Environment Rating System
NatHERS-	Nationwide House Energy Rating Scheme
NCC-	National Construction Code of Australia
NSW-	*New South Wales*
REPA-	*Resource and environmental profile analysis*
SACE-	*Selo Ambiental Colombiano para las Edificaciones*
SDGs	Sustainable development goals
TEG	Triethylene glycol
TRACI	Tool for the Reduction and Assessment of Chemical and other environmental Impacts
UN	United Nations
USA	United States of America
USGBC	United States Green Building Council

Abstract

To fulfill the increasing need for sustainable development, the construction sector seems to orient itself toward green designing. The sustainability approach is expected to generate environmental, economic, and social benefits. However, the green building design has posed consistent challenges to the reduction of the detrimental impacts of a building on the environment, as well as on human health and well-being.

Many green building rating tools have been developed worldwide to support sustainable development. These include Leadership in Energy and Environmental Design (LEED) by the United States Green Building Council (USGBC), Building Research Establishment Environmental Assessment Method (BREEAM) by the Building Research Establishment, Comprehensive Assessment System for Building Environmental Efficiency (CASBEE) by the Japanese Sustainable Building Consortium, and Green Star Environmental Rating System by the Green Building Council of Australia.

Australia is one of the top ten greenhouse gas (GHG)-emitting countries in the world and ranks second in GHG emissions released per capita. Hence, this study was conducted with the aim to support sustainable development in the country's construction industry. The Green Star environmental rating system is currently one of the most active and popular green building rating systems in the country.

Life-cycle assessment, life-cycle energy consumption, and life-cycle greenhouse gas emissions form the three pillars of life-cycle studies, which have been used to evaluate environmental impacts of building construction. Assessment of the life-cycle greenhouse gas emissions of buildings is one of the significant obstacles in evaluating green building performance. Nevertheless, in contrast, many scientists claim that there is a lack of inventory data in the construction field regarding greenhouse gas emissions. Therefore, life-cycle greenhouse gas emissions assessment needs to be considered during the initial stages of a green building project to achieve better green building design options. Moreover, it is necessary to implement a flexible and simplified assessment of life-cycle greenhouse gas emissions of commercial buildings in Australia. To assist practitioners in achieving a comprehensive overview of sustainable projects, this book develops a model to evaluate and assess life-cycle greenhouse gas emissions based on typical Australian commercial building's design options.

Further, this book explains the methodology for achieving points for the categories associated with reduction of greenhouse gas emissions in the Green Star rating system. The model for the assessment uses GaBi 8.7 platform along with Visual Basic in Microsoft Excel and shows the relationship between the building's energy consumption and greenhouse gas emissions released during the lifetime of the building. The study also illustrates that the green building design and specifications are becoming more popular and are being increasingly utilized in Australia.

1 Introduction

1.1 Background

The negative impact of climate change is an indisputable issue in the global context (Neil Adger et al., 2005; Rehan & Nehdi, 2005; Damtoft et al., 2008). One of the most significant causes of global warming is the increase of greenhouse gas (GHG) emissions into the atmosphere (Cheung, 2013; Wang & Wang, 2015; Villoria-Saez et al., 2016; Álvarez-Herránz et al., 2017). Anthropogenic GHG emissions comprise approximately 82% carbon dioxide (CO2), 9% methane (CH4), 6% nitrous oxide (N2O), and 3% other fluorinated gases (Crowley, 2000; Intergovernmental Panel on Climate Change (IPCC), 2014).

To protect the global environment, several countries with high GHG emissions have undertaken measures to improve energy efficiency, which might help reverse the uptrend of global GHG emissions since 2012. The methods for the reduction of GHG emissions should be continuously implemented via both legal and technical approaches to enhance the synchronous effects of stalling climate change and austerity (Olivier et al., 2016). To deal with the tremendous impacts of rapid global warming, governments should synchronize laws, standards, and tools to protect the environment against public nuisances as well as private interference (Wong et al., 2012; Cheung, 2013; Percival et al., 2017). One of the reasons for the rising trend of GHG emissions is the boom of global economy, to which the construction industry has been contributing significantly (Van Vuuren et al., 2017).

Green buildings have recently gained popularity due to many reasons. Numerous studies exist that focus on a variety of aspects of green buildings. However, according to Cassidy et al. (2003), green buildings are defined as buildings that have an increased usage efficiency of energy, water, and materials, such as concrete and steel. Moreover, green buildings have reduced adverse impacts on human health and environment through better planning, design, construction, operation, maintenance, and demolition throughout the complete life cycle of the building. Many studies provide similar definitions for green buildings (ASHRAE, 2006 p. 4; United States Green Building Council, 2007; Hoffman & Henn, 2008; Robichaud & Anantatmula, 2011; United States Environment Protection Agency, 2014; World Wildlife Fund, 2015). Considering all these requirements for green buildings, evaluating them has been a consistently challenging task.

Identifying the extent to which green buildings cater to these requirements is central to their evaluation and has led to the development of many green building rating tools overtime. Green building rating tools assess buildings and act as a reliable measure in the evaluation of the building for sustainability (Eichholtz et al., 2010). Green building rating tools have been developed, representing many countries and regions, giving priority to country-specific requirements (Illankoon et al., 2017a). The first green building rating tools were launched in 1990 in the UK under the name of Building Research Establishment Environmental Assessment Method (BREEAM) (Building Research Establishment Environment Assessment Method, 2015). Thereafter, the most discussed and widely used green building rating tool was launched by the United States Green Building Council (USGBC) named Leadership in Energy and Environmental Design (LEED) (United States Green Building Council, 2015). In Australia, Green Star is the most widely used green building rating tool (Green Building Council Australia, 2018).

Green Star comprises a set of rating tools developed to evaluate various green initiatives. "Green Star – Design & As Built" version 1.1 is the latest Green Star rating tool used to evaluate buildings and significant refurbishments (Green Building Council Australia, 2018). There are nine key environmental criteria illustrated in the Green Star – Design & As Built version 1.1, namely, management, indoor environment quality, energy, transport, water, materials, emissions, land use and ecology, and innovation. Each of these crucial criteria has credits, illustrating specific requirements that the building under evaluation should follow, and specific credit points are attributed if the criteria are fulfilled. Once the credit requirements are achieved, the relevant number of credit points is allocated to the evaluated building. Finally, total credit points are calculated to arrive at the final score. On the basis of the final score, a certification is awarded according to the Green Star certification criteria. However, many interdependencies among these credits exist, including credits that represent different key criteria (Tam et al., 2018a). Therefore, when considering the challenges posed by green buildings, these credits should be considered collectively in decision-making.

Life-cycle GHG emissions assessment is one of the major challenges in the evaluation of green building performance (González & García Navarro, 2006; Winston, 2010; Ahn et al., 2013). However, many scientists claim that lack of sufficient studies concerning GHG emissions considerations is the primary issue, because of lack of analyzed inventory data for environmental performance in construction field (Kawai et al., 2005; Nicol & Chadès, 2017). Therefore, life-cycle GHG emissions assessment needs to be considered during the early phases of green building projects to achieve better green building design options. According to the International Organization for Standardization (2006), life-cycle GHG emissions estimation considers the building's lifespan, including material production, construction, operation, and its final phase (which includes disposal and/or recycling phases) (Al-Ghamdi & Bilec, 2016; Tam et al., 2018b).

Considering all the aforementioned factors, it is essential to demonstrate that co-dependent credits related to reduction of GHGs should be carefully studied, as these credits are expected to lead to optimal designs for green buildings. Furthermore, when analyzing these credits, it is crucial to focus on the life-cycle GHGs rather than studying environmental impacts of GHGs produced at separate stages within the building's life cycle. Therefore, this research aims to identify optimal solutions to green building designs considering life-cycle GHG emissions in terms of Green Star credits.

1.2 Research aim and objectives

The aim of this book was to adopt a life-cycle GHG emissions assessment model based on a set of Green Star credits for commercial buildings in Australia. To satisfy the goals of this research, the following objectives were identified:

1 Critical reviews of sustainable development, green building rating tools worldwide, and techniques to analyze life-cycle GHG emissions in green buildings, which include the following:

   a Reviewing sustainable development goals and green building concepts.
   b Reviewing the development of life-cycle assessment in the construction sector.
   c Reviewing green building rating tools from global and Australian perspectives.
   d Examining model technique development in life-cycle assessment.
   e Analyzing life-cycle GHG emissions assessment in green buildings.

2 Reviewing and classifying the Green Star – Design & As Built rating tool and describing the methodology to develop a model for GHG emissions assessment:

   a Reviewing the Green Star design rating tool in Australia and determining the credits in this rating tool related to life-cycle GHG emissions.
   b Analyzing the process of model development for the life-cycle GHG emissions assessment.

3 Discussing environmental impacts of model parameters:

   a Discussing the environmental impact of commercial buildings' envelopes with respect to climate zones in Australia.
   b Discussing the environmental impacts of materials for green buildings.

4 Validating the life-cycle GHG emissions assessment model and determining optimal options for achieving credit points in the Green Star rating tool for Australian commercial buildings by employing case studies of buildings that were awarded the Green Star certificate.

1.3 The significance of the book

To fulfill the needs of the future, the Australian building sector seems to be advancing toward sustainable design. The Green Star Environmental Rating System is one of many green building rating systems that have been employed worldwide, and is currently the only Australian nationwide and voluntary rating system for the green construction (Tam et al., 2017b).

The significance of this research lies in the estimation and implementation of the flexible and straightforward process of life-cycle GHG emissions assessment for typical building fabrics and primary materials that constitute the structural materials of a commercial building in Australia and their adaptation with the Green Star rating tool. The research develops a model that can be conveniently modified to automate calculations for credits under Green Star to reduce the amount of GHG emissions during the building's life-cycle, inventively reducing tedious manual estimations, which have chronically plagued the Australian construction industry. The research also illustrates the proof of popularization of green building design and its specifications, which are being increasingly utilized in Australia.

1.4 Research methodology

The methodology of the research presented in this book includes four stages and is further divided into logical steps, as shown in Figure 1.1. The first stage includes literature review and determination of the research problems, aims, and objectives.

The second stage of the research involves the analysis of main associated elements and processes in the construction project, and selection of the credits related to the reduction of environmental impact in accordance with Green Star. This stage identifies the constraints and variables for the model analysis.

The third stage concentrates on the development of the life-cycle GHG emissions assessment model. The model uses GaBi 8.7 software to assess the life-cycle impact of each element in a building. Then, the unit results obtained are used to calculate the breakdown quantities from design. The results are ultimately consolidated in MS Excel and Visual Basic and employed to select optimal design options to achieve most Green Star credit points.

Finally, the fourth stage discusses environmental impact assessment and validation of the model. The proposed life-cycle GHG emissions assessment model is validated by evaluating three Green Star-granted buildings in Australia. These buildings (used for the case studies) are located in three different states in Australia. Detailed information on them is provided in Chapter 5.

1.5 Scope of the book

Book study proposes a life-cycle GHG emissions assessment model for an office building in Australia. The intended life-cycle GHG emissions assessment model

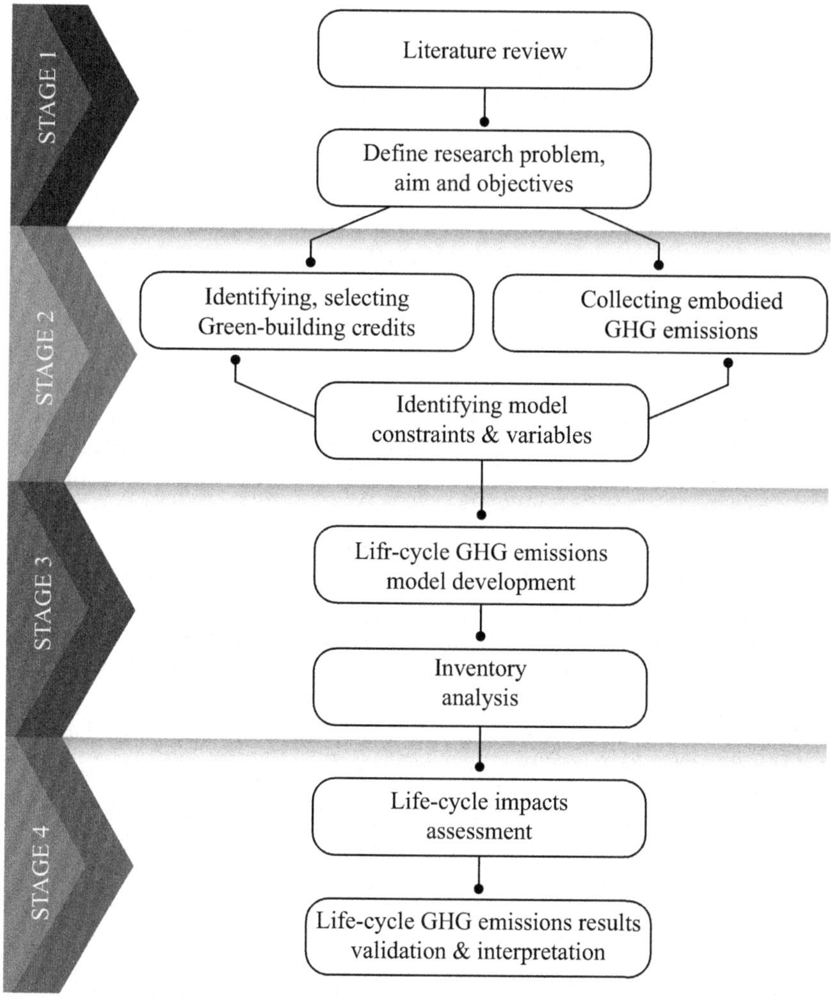

Figure 1.1 Schematic representation of the research progress

complies with the guidance of the Green Star rating tool (Design and As Built) to achieve credit points with regard to GHG emissions reduction. The environmental impact (including global warming potential, ozone depletion potential, photochemical ozone creation potential, eutrophication potential, aquatic ecotoxicity, terrestrial ecotoxicity, acidification potential, and human toxicity potential) of major materials referred in Green Star and envelope alternatives of a typical office building in Australia are assessed to find the optimal design solution associated with the lowest amount of life-cycle GHG emissions.

1.6 Book structure

Chapter 1 presents the introduction of the study. This chapter contains the research background and objectives, and a brief explanation of the research methodologies implemented.

Chapter 2 discusses the relationship between sustainable development and life-cycle assessment, as well as green building concepts in the construction industry. Furthermore, the chapter illustrates the development of green building rating systems worldwide, particularly in Australia. It also discusses GHG emissions assessment models and their limitations.

Chapter 3 focuses on the developed model to assess the environmental impacts of materials used for structures and options for building envelopes in commercial buildings in distinctive geographic climate zones in Australia. Variables of the model are described using Green Star credits, as well as features of construction projects to ensure the validity and reliability of results obtained by the model.

Chapter 4 discusses the optimal scenario in terms of Green Star credit points options that can be achieved from the model. Moreover, the environmental impact of major materials and building envelopes used in construction projects are discussed.

Chapter 5 illustrates the model validation by calculation and assessment of three case studies. These case studies, which were granted the Green Star certification, are located in three different climate zones in Australia. The validation focuses on GHG emission assessment and the Green Star credit points achieved by the projects.

Chapter 6 of this book contains the main conclusions and recommendations of the research. The chapter also presents research limitations and future directions.

2 Green building

A concept to achieve sustainable development goals

2.1 Introduction

This chapter is divided into six main sections. The first two sections of this chapter discuss the essential connection between sustainable development and life-cycle assessment, as well as the green building concept in the construction field. This section briefly presents sustainable trends in the sector and the role the life-cycle assessment plays in this growing trend. Thereafter, the chapter analyzes the development of the green building rating tools from global and Australian perspectives. The description of each of the identified credits related to greenhouse gas emission reduction in a building's lifetime is likewise included in this section. Moreover, the chapter discusses the models employed to evaluate greenhouse gas emissions of green buildings and their limitations.

This chapter also explains the procedure of life-cycle greenhouse gas emissions assessment in green building projects. First, the book examines the individual processes in the construction work and performs inventory analysis of green buildings. The next section discusses the concept of life-cycle assessment and environmental impact analysis for green buildings. Finally, a list of relevant greenhouse gas emissions assessment models is discussed to find the optimal features to apply in this book.

2.2 Sustainable development goals and green building concepts

The awareness about sustainable development was raised in the report of the World Commission on Environment and Development in 1987, which states that "Humanity has the ability to make development sustainable – to ensure that it meets the needs of the present without compromising the ability of future generations to meet their own needs" (Brundtland, 1987). Sustainable development goals (SDGs) are set in the 2030 Agenda for Sustainable Development by the General Assembly of the United Nations in 2015 (General Assembly, 2015). Seventeen goals are set for making the environment liveable, resilient, and sustainable for human beings. These SDGs replace the previous Millennium Development Goals (MDGs) to reinforce the integration of three sustainable pillars, which are the social, economic, and environmental development (Biermann et al., 2017). However, there are also many challenges to keep creating wealth and prosperity for society while protecting environmental resources.

First, the correlation between the international legislation system and national laws is not strong, as no legal obligations exist to bind governments to commit to SDGs by their national law and regulations (Kim, 2016). This is one of the SDGs characteristics that are different from other environmental goals, such as protection of the ozone layer, which are preserved in legal binders. The other distinguished feature of SDGs as opposed to previous MDGs is that they address both industrialized and developing countries (Sachs, 2012). The aim of sustainability puts all developed countries back to the level of developing countries, because of the necessity to reform their structure with a more sustainable development approach.

Moreover, SDGs, unlike MDGs, are not only related to development. In the manner of sustainable development, three major development elements, social, economic, and environmental, are adopted in these goals. These elements need to be synchronized and harmonized by governments with the international and national legislation systems. Although 169 targets are set to implement these SDGs, many of them are qualitative and give sufficient room for governments to pursue the goals on their own terms. Moreover, even when the targets are quantitative, their non-binding character still allows governments to choose the manner of interpretation and implementation of these goals.

Many activities in the construction industry have environmental impacts in the form of greenhouse gas emissions, construction waste from resource utilization, and energy consumption during phases of production, construction, demolition, and recycling of building products. These activities are directly or indirectly related to the achievement of SDGs including the health of project workers and building occupants, as well as water and energy consumption goals. The fossil fuels are still the primary energy source in the building industry, which threaten the global environment by polluting air, water, and food systems (Allen et al., 2017). The concept of a green building includes using technological innovation and effective resource utilization to decrease energy consumption and polluting emissions by reducing materials and water consumption during a building's life-cycle. This concept intersects 11 of the 17 SDGs, which are represented in Figure 2.1. The green building definition covers all concerns of building design during its life-cycle in terms of indoor comfort, responsibility of resource consumption and production, sustainable economic growth, and environmental health.

The fundamental aspects of indoor comfort are interpreted by the quality of ventilation air, water, acoustics, and lighting. The green building aims to improve the occupant's health and well-being, which is represented by SDG 3: Good health and well-being. However, the information related to these features is not sufficiently explored to date, and indoor health needs some innovation to improve the comfort level. Many green building standards such as Leadership in Energy and Environmental Design (LEED), Building Research Establishment Environmental Assessment Method (BREEAM), or the Green Star focus on improving the building's indoor comfort level as well as the health of the

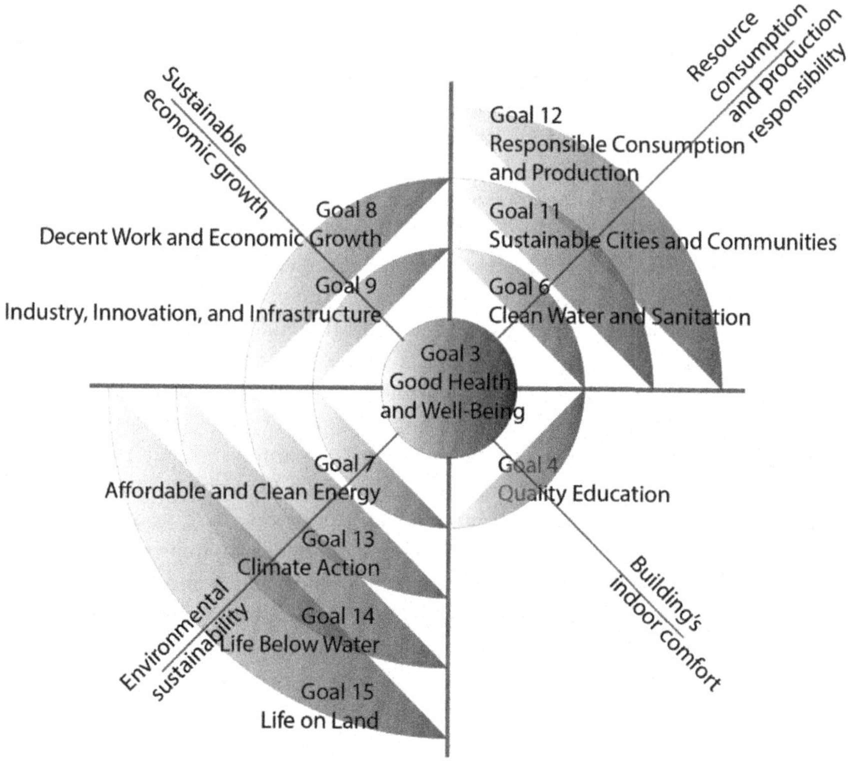

Figure 2.1 Schematic diagram of the intersection of goals of the green building concept and SDGs

occupants. With the aim of achieving SDG 4: Quality education, public awareness of green development can be raised, and practitioners can optimize the green building characteristics of their projects based on these sustainable frameworks.

Emission footprints mostly stem from the construction and manufacturing areas, which raises the necessity of materials use, resources, and energy consumption management during the building's lifetime. Regarding the targets in SDG 6, 11, and 12, the green building concept also indicates sustainable resource deployment, clean water, and hygiene management, which consolidates sustainable cities and communities. From the resulting social sustainability, economic development can be fostered by creating a stable labor market and employing innovative technologies in the industry to meet SDG 8 and 9, which encourage sustainable economic growth. In this manner, both social and economic targets combine with environmental sustainability goals to protect habitats underwater as well as on land and address the impacts of climate change.

On the basis of these SDGs criteria, many definitions of green building have been developed. However, they all have the similar aim that this practice applies incentives and processes to mitigate significant impact on the economy, society, and environment during the lifetime of the project (Bon-Gang & See, 2012; Olubunmi et al., 2016). Throughout the lifetime of a green building, the efficiency of material production, construction site operation, water, and energy utilization is improved. Furthermore, the impact on the building environment and on human physical and mental health is reduced by more effective design, construction, operation, destruction, and recycling processes.

One of the standard definitions of the green building stems from the US Environmental Protection Agency (2015b). The green building concept recognizes the initiatives of building structures and utilizes efficient processes in all phases of a building's lifetime. The concept also addresses the economic, comfort-related, and durability concerns related to the conventional building design. By this definition, a green building is recognized as a sustainable and/or high-performance building. The American Society of Heating Refrigerating and Air-Conditioning Engineers (2006) defined a green building as a project that operates with high performance considering the building's life-cycle. Natural resource utilization is minimized and alternated by renewable construction materials, water, land, and energy resources. Reducing the impact of emissions on these resources relates directly to altering global warming and climate change, or to improving the quality of the building environment including the indoor health and occupant comfort. Moreover, green design also considers the demolition and recycling phases in the life of a building. Solid and liquid waste management plans are included in the green building design to minimize adverse effects from the project. Nevertheless, this definition discusses the fundamentals of the entire lifetime and whole-building analysis in terms of environmental issues, but does not emphasize economic and social matters.

The Green Building Council of Australia identifies a green building as a building that has reduced environmental impacts during its life-cycle. By this definition, green buildings promote the building efficiency to enhance better performance and reduce its life-cycle greenhouse gas emissions significantly. Efficient resource utilization and development of a healthy environment are some significant aspects needed for the construction of a sustainable place for living and working. The sustainable indicators of a green building are illustrated in Table 2.1, representing environmental, economic, and social indicators. These indicators are categorized such that they meet the needs of sustainable development. While environmental indicators include all concerns related to human well-being, natural resources, and the environment, the social and economic indicators include population density, quality of human habitat, and macro- and microeconomic aspects (Henderson, 1994; Briassoulis, 2001; Dammann & Elle, 2006; Lawn, 2006; United Nations, 2007).

Green buildings continuously offer sufficient benefits to its occupants and relevant parties. Community recognition and perceived benefits of green buildings are becoming popular (Mahbub et al., 2012). The economic and social benefits can be recognized by the global trend that helps increase the efficiency of employment markets, strengthening the economy (Buckley & Logan, 2016).

Table 2.1 Sustainable indicators of green buildings

Social category	*Economic category*	*Environmental category*
Population, density, growth rate	Gross domestic product per capita	Global warming potential
Urban/rural migration	Investment share over GDP	Acidification potential
Accessible public services percentage	Inflation rate	Nutrient pollution potential
Quality of living environment (acoustic, thermal, lighting comfort, security)	Labor productivity and unit labor costs	Photochemical ozone formation potential
Efficient living space	The diversity of casual and permanent employment	Human toxicity potential
Lifespan and mortality rate	The adaptable capacity of the building	Depletion of fuel and heavy metal resources
	Life-cycle cost	Waste disposal and recycling

Sustainable practices and a community sense toward the green building development are likewise essential reasons for social benefits. The occupants' health and work performance benefit from the improvement of the green building's environment. In a better indoor environment, more positive social environment results are developed (Allen et al., 2017). From an economic aspect, the green building is also expected to create more jobs and save more water and energy. These lead to lower construction and operation costs for tenants and also increase asset value for building investors (Hamilton, 2015; Molenbroek et al., 2015). The most apparent benefit of the green building is its positive impact on nature. Greenhouse gas emission reduction is projected to 84 gigatons of CO2 equivalents by 2050 by saving more than 50% of energy through measures like energy efficiency and utilization of renewable energy (Dean et al., 2016). The report of the Green Building Council of Australia (2013) shows that certified green buildings help lower greenhouse gas emissions by 62% in comparison with typical buildings.

2.3 Life-cycle assessment in the construction sector

2.3.1 *Framework for life-cycle assessment*

The concept of life-cycle research has developed over decades, primarily since the 1970s and 1980s. The academic expression of "life-cycle assessment" has been embraced to reflect sustainable development studies. Furthermore, life-cycle analyses have concentrated on the measurement of energy and materials used, and emissions produced into the ecosystem during the life-cycle (Sharma et al., 2011). The life-cycle assessment, also known as life-cycle analysis, represents the procedure of assessing the potential environmental impacts that an activity or a product has over its entire life-cycle.

There are many practices that demonstrate the standards regarding life-cycle assessment, such as the Z-760 environmental life-cycle assessment guideline, published by Canadian Standards Association in 1994, or the EN 13437:2003 "Terminology, symbols and criteria for life-cycle analysis of packaging", published by the European Committee for Standardization (Menke et al., 1996; Briassoulis et al., 2010). The most realizable environmental standards are published by the International Organization for Standardization (ISO), which include ISO 14040 Environmental management, life-cycle assessment, principles and framework; ISO 14041 Environmental management, life-cycle assessment, goal definition and inventory analysis; ISO 14042 Environmental management, life-cycle assessment, life-cycle impact assessment; and ISO 14043 Environmental management, life-cycle assessment, and life-cycle interpretation (Khasreen et al., 2009).

The ISO has described life-cycle assessment as a methodology for assessing the environmental aspects and potential impact associated with a product by collecting a product system's inventory of relevant input-output, estimating the potential environmental impacts connected with these inventory data, and interpreting the inventory analysis calculations associated with the book objectives (International Organization for Standardization, 2006). Researchers suggest that life-cycle assessment should be considered as a systematic tool to estimate construction processes and elements by assessing the sequence of materials, energy use, and demolition waste, which are exposed to the environment, and studying suitable alternatives (Guinée, 2001; Cabeza et al., 2014; Abd Rashid & Yusoff, 2015).

Since 1990, many studies on life-cycle analysis have been conducted to cover issues in the construction industry as the eco-green, and sustainable trends influence all participants in the field (Taborianski & Prado, 2004; Ortiz et al., 2009). Since then, the vital function of achieving green building information by life-cycle assessment is broadly recognized, and this tool is used to recognize and implement green building practices. However, the studies in this sector still need to elaborate information on the environmental impacts of each phase in the building's life-cycle such as material production, or destruction and recycling. Besides, the life-cycle assessment for one project cannot be applied to other buildings due to differences in the building design, financial situation, and social characteristics (Singh et al., 2010).

Two fundamental characteristics of life-cycle assessment are the "cradle-to-grave" analysis approach and the use of a functional unit for relevant research (Kloepffer, 2008). Figure 2.2 shows the critical aspect of the ISO 14040:2006 standard explaining the framework for life-cycle assessment in four phases, namely: goal and scope description, inventory analysis, evaluation of life-cycle impacts, and explanation of results (International Organization for Standardization, 2006). The International Organization for Standardization (2006) indicates that the goals of the life-cycle assessment include the intended program, the book purpose, and a targeted audience. The goal and scope definition is the first stage of the life-cycle assessment, and identifies the research objectives, system boundary, functional unit, and other limitations. The functional unit in this

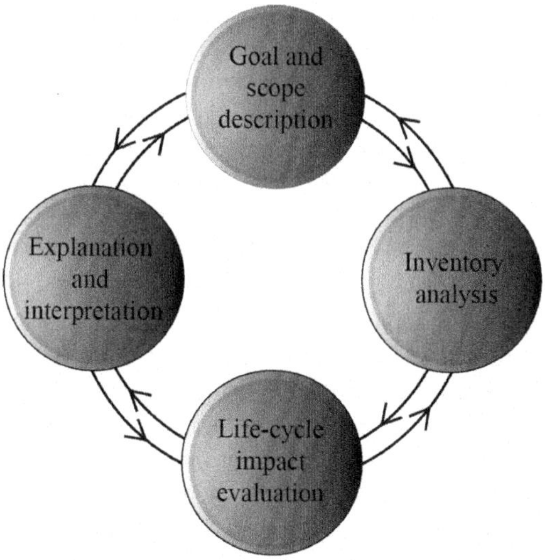

Figure 2.2 ISO 14040:2006 Schematic representation of the life-cycle assessment framework

phase delivers the definition of a reference to which all the data are associated. In the building industry, functional units include as many compulsory properties as can be studied (Weidema et al., 2004). This stage determines the information category that is necessary to lead to the final decisions, the accuracy level, and the interpretation and presentation method for the results. The principal goal is choosing the optimal product, process, or service with the minimum impact on the indoor and outdoor environment. Taking a life-cycle assessment can also help guide the innovative development toward a net decline in energy consumption and emissions. There might also be subordinate goals for executing a life-cycle assessment, which would change according to the building type (Curran, 2006).

2.3.2 Life-cycle inventory and life-cycle impact assessment

Life-cycle inventory (LCI) analysis, which is the second phase of life-cycle assessment, involves gathering data for unit processes including all related input and output flows of energy and other resources, as well as the data of pollution emission to the environment throughout the building's life-cycle. This phase comprises estimating quantities of materials, as well as inputs and outputs of energy consumption, in a construction system (Klöpffer & Grahl, 2014).

Along with the LCI analysis phase, life-cycle impact assessment evaluates the consumed resources and their potential impact on the environment. Three necessary elements in this phase include combinations of impact categories, classifications of LCI results, and characterization of category indicators (Klöpffer & Grahl,

2014). Classification of LCI results is followed by assignment of impact categories along with category indicators. The category indicator could be analyzed at any step between the LCI results and the category endpoint, where the environmental impacts occur (Jolliet et al., 2003). There are two fundamental approaches in a life-cycle impact assessment: problem-oriented approaches (midpoint methods) and damage-oriented approaches (endpoint methods) (Bare et al., 2000). In the problem-oriented approaches, the environmental involvement from some phases in the LCI are analyzed along with the ultimate damage caused by this involvement (Azapagic, 2006). On the other hand, damage-oriented approaches are commonly referred to as "endpoint" methods because they assess the environmental damage caused by the interferences to human habitats (Bare et al., 2000).

Several methods are classified according to their analysis of thermal transmission through building elements for the development of building energy simulation models. These include response factor methods, conduction transfer functions, finite difference methods, and lumped capacitance models. Preferably, these are used in combination, in order to increase the strength of each method, rather than separately (Kramer et al., 2012; Harish & Kumar, 2014).

Room thermal response factors are a convenient way of depicting data of the dynamic thermal characteristics of a room for the modelling of air conditioning, which was formally introduced by Mitalas and Stephenson (1967). A time series of responses of the room cooling load and surface temperatures is calculated from a time series of unit pulses by solving the heat balance equations for all the surfaces and the ambient air (Mitalas & Stephenson, 1967; Kramer et al., 2012).

The response of factors for a particular room's components, which contain the relevant information about the room including its size, construction type, color, and surface emissivity, can be combined with the appropriate excitation to obtain the response in a specific case (Stephenson & Mitalas, 1967). Conduction transfer functions, such as Laplace transform method, state-space method, direct root finding method, and frequency-domain regression method, are efficient methods to compute surface heat fluxes because they do not require data on temperature and fluxes within the surface (Stephenson & Mitalas, 1967, 1971).

The finite difference method is a well-established and conceptually simple method for solving heat transfer problems (Ozisik, 1994), which requires a pointwise approximation to the governing equations (Lewis et al., 2004). In this method, a wall is divided into a finite number of control volumes for which the heat balance equations are solved (Kramer et al., 2012). However, the finite difference technique becomes cumbersome when unusual geometries are encountered (Lewis et al., 2004).

The lumped capacitance model, also known as lumped system analysis (Incropera, 2011), can be used to solve transient heating and cooling problems by reducing a thermal system to a number of discrete "lumps" and assuming that the temperature difference inside each lump is negligible (Virag et al., 2011). The electrical networks, made up of resistances and capacitances, are used to model the building element. The capacitance represents thermal capacitance of (a layer of) a wall (Kramer et al., 2012).

In terms of parameter identification techniques, physical or white-box (sometimes called glass-box) modelling (Brause, 2010), statistical or black-box modelling, and physical statistical or gray modelling are used to estimate the building's energy consumption (Mustafaraj et al., 2011; Amara et al., 2015). As shown in Figure 2.3, while a detailed and structured amount of internal information about a building is required to apply the white-box method, the basic aspects of the system are investigated against both internal and external factors of the studied building.

2.3.3 Life-cycle assessment methodologies

Life-cycle assessment methodologies in the construction sector are still in a developing stage because of the discrepancy and diversity of buildings with respect to material production and utilization, site location, construction methods, and technologies used during the project's lifetime. All of these aspects will lead to distinctive goals and scope of life-cycle assessment. Many modern life-cycle assessment methodologies are developed, such as CML, ILCD, IMPACT, TRACI, and ReCiPe. The application of "life-cycle impact assessment" emerged in the early 1970s with the technical term "resource and environmental profile analysis (REPA)". The first life-cycle assessment method was deployed in 1992 as an environmental assessment standard, termed CML 1992. The International Reference Life Cycle Data System (ILCD) has been modelled by the Joint Research Centre – Institute for Environment and Sustainability in European

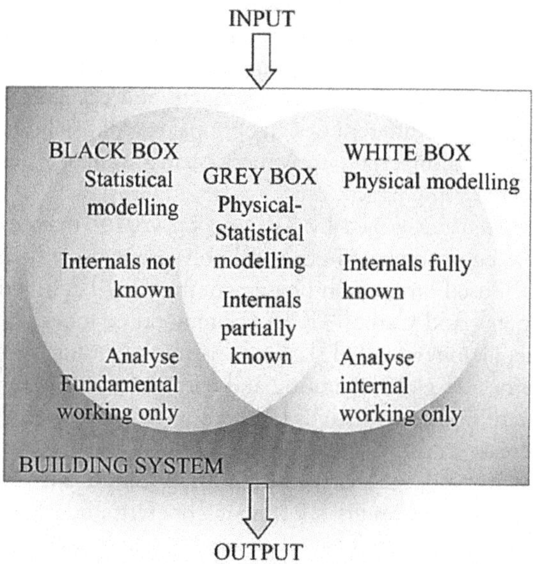

Figure 2.3 Modelling categories in terms of parameter identification techniques

Commission, to deliver guidance on coherent and quality-assured life-cycle assessment data and studies (European Commission, 2010). ILCD is a handbook that complies with the ISO 14040/44 standards. It provides LCI datasets that best imitate situations associated with energy consumption and greenhouse gas emissions production.

IMPACT 2002+ methodology, which was established by the Swiss Federal Institute of Technology – Lausanne (EPFL), restricts quantitative modelling to limit uncertainties and groups LCI results in so-called midpoint categories (Jolliet et al., 2003). The human, aquatic, and terrestrial toxicity factors have been taken from this methodology (Frischknecht et al., 2007).

TRACI (Tool for the Reduction and Assessment of Chemical and other environmental Impacts) also allows practitioners to assess potential environmental effects. This tool has been developed by the United States Environmental Protection Agency (Bare et al., 2012). ReCiPe 2016 is the latest version of the life-cycle assessment tool series, which has been upgraded from ReCiPe in 2008. This tool assesses different environmental effects and has been developed by RIVM, Radboud University Nijmegen, Leiden University, and Pré Consultants (Huijbregts et al., 2017).

There are two remarkable life-cycle assessment tools, GaBi and SimaPro, that are able to assess life-cycle greenhouse gas emissions during a building's life-cycle (Zhang & Wang, 2015). Both programs can use a variety of eco-datasets to flexibly evaluate the life-cycle impacts of construction projects. Ecoinvent, U.S.LCI, and AusLCI databases, which support systematic data for construction projects, can be coordinated in GaBi as well as SimaPro.

GaBi has some advantages over SimaPro such as developing a model with multiple suppliers in its datasets and ability to connect between different products from different suppliers (Rodríguez & Ciroth, 2016; Tam et al., 2018a). GaBi is a life-cycle assessment program for practitioners with limited capabilities and with a variety of outputs from life-cycle assessment datasets (Tam et al., 2018a). Using GaBi software, project planners can investigate and compare released greenhouse gas emissions from different design options regarding a building's life-cycle, including material manufacture, construction, maintenance and replacement, and disposal and recycling phases.

A model has been developed by Kua and Lu (2016) to assess the environmental performance of tempered-glass and polycarbonate windows. The model deployed by GaBi used production phase data from the Ecoinvent database and Inventory of Energy and Carbon (ICE) (Hammond & Jones, 2008). The book suggested that replacing tempered glass in construction projects with polycarbonate will reduce the net global warming and eutrophication potential. However, other environmental impacts such as human toxicity potential and aquatic ecotoxicity will increase in turn.

Another model was developed within GaBi by Sim et al. (2016) to assess a 25-storey building. Their results show that concrete and steel have negative impacts on the environment. They also point out that cement and glass products contribute significantly to the ozone depletion potential. Mangan and Oral (2015) developed a model to assess life-cycle energy consumption and life-cycle greenhouse gas emissions for refurbishment options upgrading existing dwelling energy systems in Turkey using GaBi with ICE data. Their study showed that retrofit strategies and climate zones represent two aspects that have significant impacts on life-cycle energy consumption and life-cycle greenhouse gas emissions results.

Bueno et al. (2016) analyzed external wall structure systems for buildings in Brazil using GaBi. In this book, the characterization models were affected by regional and chronological matters. The study found that the Impact World+ method is capable of providing evaluation modules for many regions. However, their study also highlighted that practical life-cycle impact assessment tools for accessing maximum global environmental data are still lacking. The authors implied that practitioners should deploy more than one life-cycle impact assessment method to select the best project designs based on the comparative results of these methods.

Different algorithms predicting building energy consumption can vary from modelling of a single component (a slab, a wall, etc.) to modelling of a whole building in different climate conditions (Harish & Kumar, 2016). These computational models are used to support sustainable design analysis, which provide rapid and quantifiable feedback on several sustainable alternatives and answers to "what if" questions posed by the design team and client in the early planning stage (Emmitt, 2013).

A vast library of life-cycle inventories database for over 60 countries, including the Australian National Life Cycle Inventory Database (AusLCI) dataset, can be employed in GaBi (Tam et al., 2018a). The model proposed in this book uses the AusLCI database to implement the analysis. AusLCI is a significant initiative currently being developed by the Commonwealth Scientific and Industrial Research Organisation (CSIRO) and the Australian Life Cycle Assessment Society (ALCAS) (Giurco et al., 2008).

The sustainability challenges in cities can be overcome in ways that allow them to continue to thrive and grow, while improving the use of resources and mitigating pollution and poverty. The future we strive for includes cities full of opportunities for all, with access to essential services, energy, housing, transportation, and more. The life-cycle cost, life-cycle energy consumption, and life-cycle greenhouse gas emissions assessments form three pillars of life-cycle studies that have been used to evaluate the environmental impact in building construction (Aïtcin & Mindess, 2011; Chau et al., 2015; Tam et al., 2017a).

2.4 Green building rating tools: global and Australian perspectives

The life-cycle assessment of a building is a technique applied to evaluate the environmental impacts throughout the building's lifespan. Guidelines for achieving life-cycle assessment results are provided by the ISO via ISO 14040 (2006) and ISO 14044 (2006). Figure 2.4 shows the analytical procedure of life-cycle assessment, which evaluates all the resource inputs of a product including the energy consumption, water and materials, and greenhouse gas emissions, as well

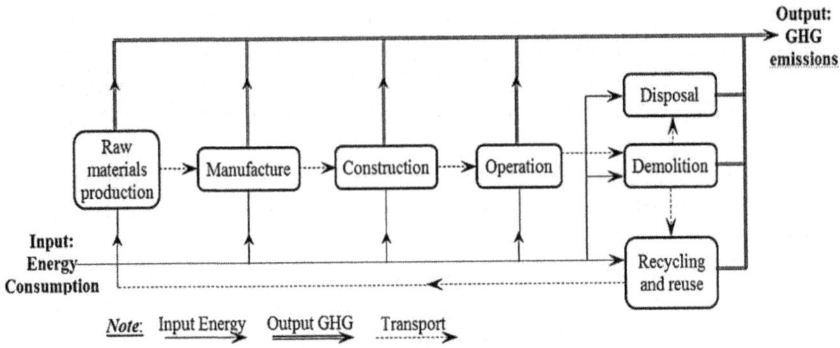

Figure 2.4 Theoretical process of life-cycle assessment

as solid and liquid wastes, at all stages of production, construction, use, and the end-of-life stage of the building. While life-cycle energy analysis is an approach that accounts for all energy inputs to a building in its life-cycle, greenhouse gas emissions are assessed as the construction output to the environment during the lifetime of the building (Ramesh et al., 2010; Chau et al., 2015).

Anthropogenic global climate change is likely an inescapable truth for all living species on the Earth (Intergovernmental Panel on Climate Change (IPCC), 2007). This is due to the significant accumulation of activities in worldwide construction sectors, leading to rapid upsurge of greenhouse gas emissions, including carbon monoxide (CO), carbon dioxide (CO2), methane (CH4), and other greenhouse gas concentrations (Omer, 2008; Hong et al., 2015; Holmgren et al., 2017). In 2016, the atmospheric CO2 level surged to the highest record observed in the last millennium (World Meteorological Organization (WMO), 2017). In particular, the construction sector has accounted for approximately 20% of the total global energy consumption, and this energy demand is projected to continuously increase at an average rate of 1.4%/year in residential areas, and approximately 1.6%/year in service areas during 2012–2040 (U.S. Energy Information Administration, 2016).

Several international green building rating systems consist of life-cycle assessment (Al-Ghamdi & Bilec, 2016; Illankoon et al., 2017b). There are 27 countries around the world, classified into five major regions, with established green building councils (Table 2.2), which include LEED by United States Green Building Council, California Green Building Standards (CALGreen) by California Building Standards Commission, BREEAM by UK Building Research Establishment, Green Globes by Green Building Initiative, Comprehensive Assessment System for Building Environmental Efficiency (CASBEE) by Japan Sustainable Building Consortium, the Green Star by Green Building Council of Australia, Building Environmental Assessment Method (BEAM Plus) by Hong Kong Green Building Council, Green Mark by Building Construction Authority Singapore, and International Green Construction Code (IgCC) by International Code Council (ICC).

Table 2.2 Current green building rating tools around the globe

Region	Country	Name of the green building council (GBC)	Green building rating tool
Africa	South Africa	GBC South Africa	The Green Star SA (adapted from Green Star Australia)
America	Argentina	Argentina GBC	LEED
	Brazil	GBC Brazil	LEED, AQUA (Alta Qualidade Ambiental)
	Canada	Canada GBC	LEED Canada, Green Globes
	Chile	Chile GBC	LEED
	Columbia	Columbia GBC	Sello Ambiental Colombiano para las Edificaciones (SACE), [Colombian Environmental Seal for Sustainable Building]
	Peru	Peru GBC	LEED
	USA	US GBC	LEED, Green Globes
Asia-Pacific	Australia	GBC Australia	Green Star
	Chinese Taipei	Taiwan GBC	EEWS (Ecology, Energy saving, Waste and Health), LEED
	New Zealand	New Zealand GBC	The Green Star adapted from Green Star Australia)
	Hong Kong	Hong Kong GBC	BEAM Plus (Building Environmental Assessment Method)
	India	Indian GBC	IGBC (Indian Green Building Council) Rating, LEED
	Japan	Japan Sustainable Building Consortium	CASBEE (Comprehensive Assessment System Built Environment Efficiency)
	Malaysia	Malaysia Green Building Confederation	Green Building Index
	Singapore	Singapore GBC	Green Mark
Europe	Croatia	Croatia GBC	LEED, BREEAM, DGNB (Deutsche Gesellschaft für Nachhaltiges Bauen)
	France	France GBC	HQE (High Environmental Quality)
	Germany	German Sustainable Building Council	DGNB, BREEAM Germany
	Netherlands	Dutch GBC	BREEAM
	Poland	Polish GBC	LEED, BREEAM
	Spain	GBC Espana	VERDE (Green) (From SB Tool), BREEAM, LEED
	Sweden	Sweden GBC	BREEAM SE (Swedish version), LEED
	Turkey	Turkish GBC	BREEAM, LEED, DGNB
	UK	UK GBC	BREEAM
Middle East and North Africa	UAE	Emirates GBC	LEED, BREEAM, ESTIDAMA
	Jordan	Jordan GBC	LEED

In terms of the green building rating tool utilization, BREEAM and LEED are widely represented in the American, European, and North African regions. Moreover, these tools were used to form other tools such as the Green Star in Australia or HK-BEAM in Hong Kong (Ding, 2008; Alyami & Rezgui, 2012; Tam et al., 2017c). In accordance with the ISO 14040 series, a "cradle-to-grave" assessment for a building should be implemented to improve the building's environment friendly aspects (International Organization for Standardization, 2006).

Furthermore, a crucial phase in the building's lifetime is the operational phase, which can be assessed by "Energy" criteria. This is illustrated in Table 2.3, where the "Energy" criteria have been found to have the highest weightings on all the green building rating tools except for CASBEE (Villoria-Saez et al., 2016; Illankoon et al., 2017b; Tam et al., 2017b).

2.5 Model technique development for life-cycle assessment

In the efforts of efficient life-cycle assessment of a building, a vast diversity of building performance programs has been developed. A summary of conventional energy simulation tools is shown in Table 2.4. Energy efficiency designs including the building envelope, building orientation, and natural ventilation lead to a decrease of dependence on energy-consuming mechanical systems in maintaining comfortable indoor temperatures of a building (Zhu et al., 2013). Addressing energy consumption during the lifetime of a building involves not only the energy demand in the operation phase but also the material production, construction, and demolishment, as well as recycling stages (Yuan & Jin, 2015). Therefore, the more practical the building, the more complicated the model that the experts need to develop in order to precisely access the building's environmental impacts.

Apart from numerous tools and plug-ins for model development in building energy simulation, a case study by Crawley et al. (2008) has presented features of 20 major applications that are commonly used for building energy simulation, namely: BLAST, BSim, DeST, DOE-2.1E, ECOTECT, EnerWin, Energy Express, Energy-10, EnergyPlus, eQUEST, ESP-r, IDA ICE, IES/VES, HAP, HEED, PowerDomus, SUNREL, Tas, TRACE, and TRNSYS. Since then, hundreds of programs have been developed, containing many methods for estimation of energy

Table 2.3 Energy criteria weighting in international green building rating tools

Tool	*Energy criteria (%)*	*Maximum green building credit points*
LEED	30.43	115
BREEAM	23.08	130
Green Star	24.00	100
Green Mark	49.73	183
GBI	39.36	94
BEAM plus	34.29	140
IGBC	29.17	96
CASBEE	8.43	83

Table 2.4 Up-to-date life-cycle assessment tools

Program	*Working environment*	*Output*	*Cost*	*Developer*
IES	a1, a2, a3, a4	b1	Variable	IES
GaBi	a4	b1	Variable	Thinkstep
SimaPro	a4	b1	Variable	PRé Sustainability
EnergyPlus	a2	b1, b2	Free	US DOE
Sefaira	a1, a2	b1	Variable	Sefaira
GBS	a1, a4	b1, b2, b3	Autodesk extension	Autodesk
OpenLCA Nexus	a4	b1	Free	GreenDelta GmbH

Note: a1: Autodesk Revit; a2: SketchUp; a3: Vectorworks; a4: Other modelling programs exports to the green building XML schema (gbXML);

b1: Whole-building energy simulation; b2: Energy conservation measures; b3: Greenhouse gas emissions reporting.

performance, such as whole-building energy simulation, parametric and optimization, model input calibration, energy conservation measures, building energy auditing, life-cycle analysis, and cost analysis (Figure 2.5). This demonstrates the capabilities of 134 latest tools for simulating the energy performance of a building (Hong et al., 2000; International Building Performance Simulation Association (IBPSA), 2017). However, these high-end simulation engines, for instance, DOE-2.2 and Energy Plus, require detailed information and are time-consuming; both information and time are always in short supply during the initial stages of every project (Chowdhury et al., 2007).

Moreover, the more structured tools tend to be more straightforward in design, while the modular/flexible simulation programs require additional modelling time (Kilkelly, 2015). Since their introduction in the 1970s, building energy management systems have become one of the trends that provide comfortable indoor environments and effective utilization of energy in buildings (Perera et al., 2016). Perera et al. (2016) used both Matlab and Modelica to simulate the environment that covers multi-floor physics-based building heating models for an existing Norwegian three-storey residential building built in 1987. The Matlab model consisted of 16 state variables including the air densities of each floor, while the Modelica model illustrated the effect of wind, which was not included in the Matlab model. However, after its development using object-oriented language such as C# or Java, the model run in Matlab environment is simpler compared to Modelica, more accessible for structural configurations, and more comfortable to integrate into the building energy management system.

It is the opinion of some investors that buildings with low carbon emissions are harder to sell and less profitable, which forms a barrier to commercializing housing using new technology to build sustainable buildings. Recently, effective multi-objective optimization models will change these views and strengthen the investors' confidence in the green building endeavor (Hamdy et al., 2016; Khodabuccus & Lee, 2016). A baseline design was deployed by Khodabuccus and Lee (2016) using

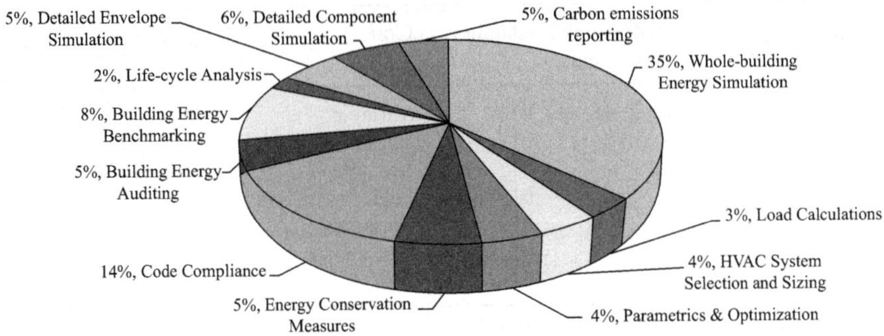

Figure 2.5 Capabilities of building energy performance simulation tools

a detached four-bedroom property to acquire a technically and economically linked model to optimize zero carbon design and develop a home using this technique. The model was run in the TRNSYS application and the building was divided into 15 individual thermal zones with different occupancy patterns and usages. Standalone tools and plug-ins that contain whole-building energy simulation capability in the building's life-cycle along with flexible methods for geometrically creating valid input files are currently being developed in order to easily integrate information into the existing building modelling tool and enable regular monitoring of building performance in the early planning stage (Kilkelly, 2015).

In life-cycle assessment studies, OpenLCA Nexus is a free software developed by GreenDelta GmbH to provide a footprint model with a large library of free and purchased databases. However, this software requires developers to have knowledge of SQL, Python, or JavaScript languages. Nevertheless, with free database options, the program is quite simple to build, given the complexity of the whole-building model, in particular, climate conditions (GreenDelta GmbH, 2017). To facilitate the estimation of energy consumption of a building, the national laboratory of the U.S. Department of Energy (2017) developed OpenStudio as an open source cross-platform that uses EnergyPlus and SketchUp to support whole-building energy modelling. Although this software is freeware, its SketchUp plug-in is required to run it in the updated SketchUp Pro version, which can be purchased via the Trimble Inc. (2017)'s website; for example, the latest version 2.2.0 of OpenStudio is only supported in SketchUp Pro 2017. Alternatively, even experts can utilize Microsoft Excel to develop their model. However, to build this kind of model, great effort is needed to develop the database as well as to analyze a large amount of data from input and output resources, which require execution by other plug-ins such as Visual Basic or other programming languages that match with Microsoft Excel (Tam et al., 2017d). One part of the project implemented in this chapter has been developed in Microsoft Excel and Visual Basic.

Developing building models for the estimation of life-cycle energy consumption, as well as life-cycle greenhouse gas emissions, is one of many contemporary approaches to address global climate change challenges. Nowadays, with the assistance of a series of computer programs, calculation of the energy used in a building is easier than ever. There are many studies associated with the development of building models for managing greenhouse gas emissions produced by the energy use in building operations. However, the results from life-cycle environmental and economic impacts assessment of the same project by different life-cycle assessment methods are not always similar and coherent (Herrmann & Moltesen, 2015). Therefore, it is hard to decide which design scenario will be more suitable for the environment as well as for the investor's financial plan. This situation also occurs in the sustainable design community in Australia. The country has pledged to decrease approximately 26% of greenhouse gas emissions by 2030 compared to the 2005 emissions level (Shahiduzzaman & Layton, 2015).

The characteristics of building materials, construction, and space types, which would be analyzed to achieve the credit points in Green Star, can be obtained from their databases or added from external sources (Green Building Council of Australia, 2015a; Autodesk Sustainability Workshop, 2017; International Building Performance Simulation Association (IBPSA), 2017). Sustainable development creates an efficient and economical approach to building cleaner and greener buildings for society. Both GaBi and SimaPro have their individual and unique databases and methods to assess energy consumption, as well as greenhouse gas emissions released during the building's lifetime. As stated in the introduction, this chapter evaluates whether these programs are suitable for use in the life-cycle assessment approach along with the Green Star rating application. The following sections present the findings of this book.

2.5.1 *Compatibility of computational assessment programs with Green Star rating tool*

The Green Building Council of Australia encourages greenhouse gas emissions reduction along with reduction in the energy consumption during building operation by five alternative pathways. The pathways in Green Star's energy credit are a testimony that the proposed building successfully complies with the requirements in the NatHERS rating tool by Nationwide House Energy Rating Scheme (2007) for dwellings located in all states and territories in Australia (with the exception of New South Wales (Credit 15B)), or in the BASIX assessment tool by NSW Government, the Building Sustainability Index (2009) for projects located in New South Wales (Credit 15C), or in the NABERS Energy Commitment Agreement by National Australian Built Environment Rating System (2013) (Credit 15D). The other pathways also require that the designated building demonstrates the reduction of its greenhouse gas emissions by employing "best practice" elements in Credit 15A or demonstrates that the greenhouse gas emissions are less than those of the reference building in Credit 15E (Green Building Council of Australia, 2015a).

According to Renouf et al. (2015) and Myhre et al. (2013), the optimal practice recommendation for life-cycle assessment in Australia is the Centre for Environmental Studies (CML) method, which is developed by the Institute of Environmental Sciences, Leiden University, the Netherlands (Guinée, 2001), as it aligns best with the Australian Environmental Product Declaration scheme's requirements. Both GaBi and SimaPro can be applied in the normalization step of the CML 2001 method. However, despite applying a similar methodology, results from these applications are still notably different due to the different normalization reference approach of this method, for example, the significantly higher factor of carbon footprint for emissions of hydrogen fluoride in the air in GaBi, or the difference in the assigned name of chromium emissions in each software (Herrmann & Moltesen, 2015).

Practitioners and designers can use both these programs to assess energy consumption and greenhouse gas emissions associated with the construction to achieve up to 34 credit points in Green Star (Green Building Council of Australia, 2015a, 2015b). Because both applications emphasize the role of life-cycle assessment in product development and provide increased sensitivity for environmental impact (Techato et al., 2009), the processes can also be customized with the project's desired parameters in the analysis (Salcido et al., 2016). The accuracy of the results will depend on the data available in the software database (Weißenberger et al., 2014; Židonienė & Kruopienė, 2015).

Figure 2.6 illustrates the aspects considered for achieving credit points from those pathways. Both GaBi and SimaPro provide options to calculate and assess not merely the building's energy consumption, but also the environmental impact owing to their ability to provide a significantly efficient estimation engine in accordance with the updated impact assessment approaches and the vast updated database of energy-related elements (building fabric, glazing, air conditioning and ventilation systems, heat water supply, etc.), which are mentioned in NCC (Australian Building Codes Board, 2013).

2.5.2 *Challenges in using computational programs for life-cycle assessment*

Although both programs have massive databases of processes, flows, and other aspects related to forming a building model, these large databases also pose challenges when choosing appropriate components for the model. Researchers have to go through this difficult process in all cases. Moreover, this information refers mostly to general procedures in Europe, mainly because both programs were first supplied in the European region: GaBi in Germany and SimaPro in the Netherlands. Even though their updated versions were repeatedly revised, they still do not adequately cover the aspects of sustainable research in particular countries. For example, in GaBi, there are about 300 default processes for Australia, while in the European region the program provides more than 600 processes.

Another limitation for creating a new building model to simulate energy and calculate greenhouse gas emissions in these two programs is the geographic location, which needs to be included in the model. In the design process, practitioners must combine these life-cycle assessment programs with other design software such as AutoCAD or Revit. The application of whole-building life-cycle assessment is complicated due to the variety of locations that the model has to consider, which leads to the difficult task of considering the data of meteorological

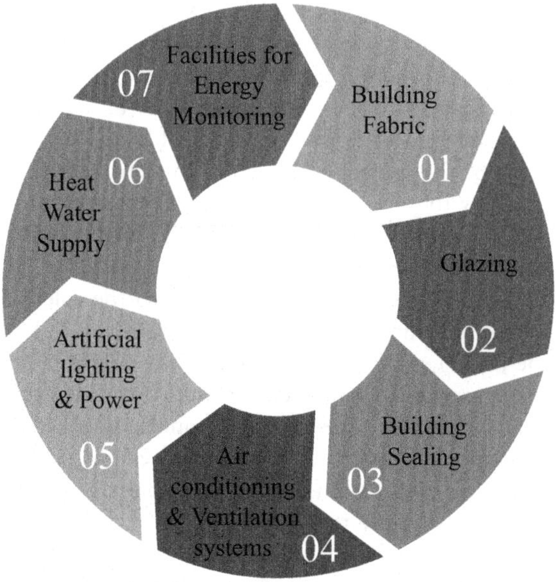

Figure 2.6 Flowchart of possible implications in computational assessment programs with Green Star

conditions of these places. In addition, life-cycle assessment results that are exported from these programs should be interpreted carefully by life-cycle assessment experts and other practitioners to prevent any bias (Herrmann & Moltesen, 2015). This is a barrier for designers and other stakeholders who are not familiar with these programs.

These two programs have trial versions, but their functions are limited and inadequate for building complicated models. The trial packages, mainly for education purposes, have few databases that allow users to understand how these programs work. Regarding the commercial version, SimaPro seems to be more expensive than GaBi for both annual subscription and perpetual license. However, GaBi divides LCI databases into many packages that need to be purchased in order to be applied to the life-cycle assessment of specific areas.

2.6 Life-cycle greenhouse gas emissions analysis in green buildings

2.6.1 Construction project process breakdown and inventory analysis

A regular construction project employs a variety of material resources, machinery, and equipment, which generate many types of air pollutants. To consider and precisely evaluate the environmental impact of a building, it is essential to break down the entire project into separate unit processes. Subsequently, the work breakdown analysis can be used in the building's life-cycle assessment. From a systematic approach, a building process can be analyzed as a collection of dynamic and inter-dependent construction items such as formwork, reinforcement, concrete, masonry, and finishes. These work items can be further investigated as an organized group of unit processes that are satisfied by the separative and technical requirements in the project plans. From this breakdown, the number of work items will satisfy the inventory analysis requirements and provide crucial information for future assessments. This represents the fundamental approach to implement the life-cycle assessment for a construction project.

The environmental impact estimation is based on the calculation of greenhouse gas emissions, including those of carbon monoxide (CO), carbon dioxide (CO2), methane (CH4), and other greenhouse gases (Omer, 2008; Hong et al., 2015). The greenhouse gas emissions released during the life-cycle of a building include the emissions from the extraction and manufacture of construction products, construction, transportation, operation, and end-of-life phases (demolition, disposal, recycling, and reuse) (Hammond & Jones, 2008; Le et al., 2018a). Nevertheless, the process of the research in life-cycle greenhouse gas emissions assessment is less certain and consistent due to assumptions that are applied in the analysis of greenhouse gas emission assessment. Although the life-cycle assessment of a construction project provides an extensive evaluation of building design alternatives, there are still some obstacles that prevent effective handling of design decisions in the early phases of the project. Moreover, the inaccuracy of environmental impacts during and after the life-cycle of a building leads to more uncertain and unreliable outcomes.

Several reasons contribute to the lack of reliable and accurate results, which include (1) industrial procedures dictate the inventory data of life-cycle greenhouse gas emissions of construction products and materials, while details of a construction project are designed by a project team including designers, engineers, and so forth; (2) system boundaries for construction material and building envelopes are based on a single product analysis, while a whole construction project is a complicated process that needs many assumptions to be adopted.

After the construction process breakdown, following the impact analysis steps of the life-cycle assessment procedure, the environmental impact of the building is evaluated at the level of unit processes and work items, as well as at the whole-building level. The inventory analysis for the assessment quantifies the process inputs, such as material and energy resources, and the output to the environment. The breakdown of a building processes should be implemented in accordance with the measurement guidelines of the Australian Institute of Quantity Surveyors & Master Builders Australia (2016) and according to the actual projects before inventory analysis.

Fundamental input data types required for inventory analysis include (1) building information, for example, building location; (2) quantity of construction material; (3) machinery information, for example, type, amount, and operational time of equipment, energy type, etc.; (4) subordinate data of input materials, such as type, amount, and recycled ratio of materials/products. These data can be obtained from building design and related documents.

2.6.2 *Life-cycle assessment framework and environmental impact analysis for green buildings*

Basic background knowledge of the methodologies of life-cycle assessment and greenhouse gas emission analysis is provided in the literature review of Chapter 2. The life-cycle assessment considers the optimal methodology and is extensively recognized by previous research in the context of life-cycle greenhouse gas emissions analysis. The review of previous studies and methodologies leads to the development of the book framework, as shown in Figure 1.1.

In the first stage of the research, the literature review focuses on life-cycle greenhouse gas emissions assessment, which leads to the definition of the research problem, the objectives, and aims of the book. The life-cycle greenhouse gas emissions estimation is one of the fundamental features of this research. This part represents the third stage of the book. In this stage, the Green Star credit classification is introduced, which relates to the reduction of greenhouse gas emissions during the life-cycle of buildings in Australia. Thus, credit classification is used to estimate the life-cycle greenhouse gas emissions based on the requirements of the Green Star rating tool. After the model development, two remaining sections include the explanation of results and their validation.

The methodology for life-cycle greenhouse gas emissions assessment is used to expand the comprehensiveness of the system boundaries from cradle-to-grave of the commercial buildings in Australia. The research aim is to quantify life-cycle greenhouse gas emissions with different building envelopes and concrete type alternatives (Figure 2.7). One of the research objectives is to develop an inventory database of greenhouse gas emissions using the life-cycle assessment method, which is extensive and comprehensive in taking into account the model's system boundaries. The system boundaries of the research are included in the lifetime of the construction project procedures.

An extensive environmental impact analysis is required to interpret all primary inventory data regarding potential environmental impacts on human health, the ecosystem, and other factors. The impact analysis comprises three subordinate steps: classification, characterization, and valuation (Vigon, 1993; Burgess & Brennan, 2001; Li et al., 2010). Input and output impact categories are determined and connected in the process of classification. Then, the valuation is implemented after characterization, that is, the LCI results calculation and conversion to common units.

2.6.3 *Life-cycle greenhouse gas emissions assessment models*

In terms of the program that can be used for the model simulation, the literature review illustrates that both GaBi and SimaPro satisfy the model's requirements with the flexible integration of a number of eco-databases, among which are the

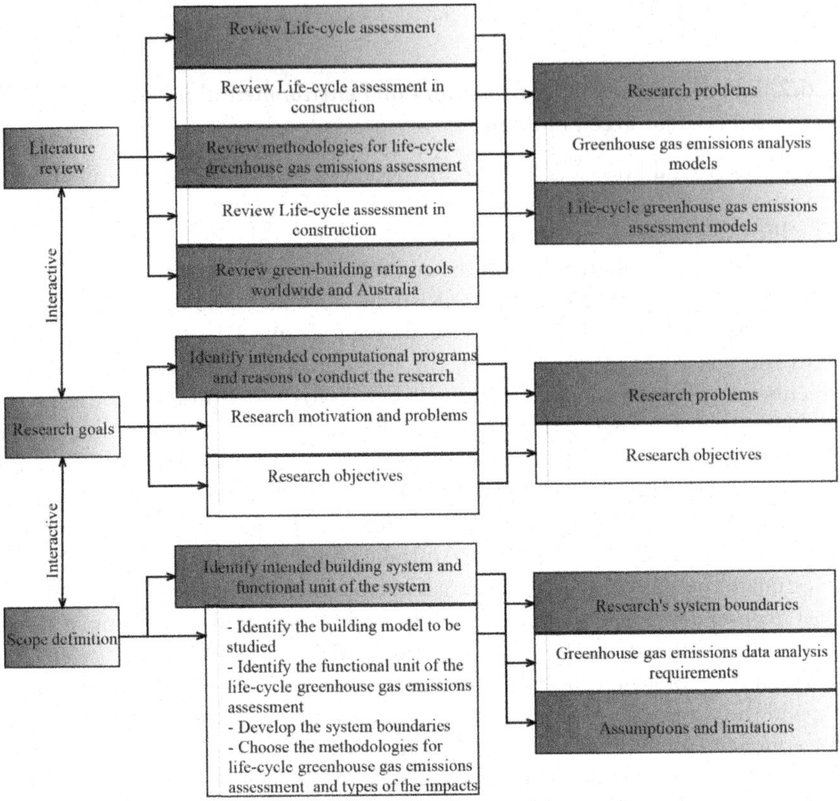

Figure 2.7 Flowchart of research aims and scope definition

Ecoinvent, U.S.LCI, and AusLCI database, that provide well-documented process data for thousands of products, assisting experts to make genuinely informed options about their eco-friendly impact (refer Table 2.5).

With the huge library of environmental impact analysis methodologies in GaBi and SimaPro, it is possible for any expert idea within the life-cycle assessment framework by using the up-to-date methods such as ReCiPe 46 and ILCD (Finnveden et al., 2009; European Commission, 2010; National Renewable Energy laboratory, 2012; Islam et al., 2015a; Adibi et al., 2017; Australian Life-cycle Assessment Society (ALCAS), 2017; Ecoinvent, 2017; GaBi, 2017a; Huijbregts et al., 2017; SimaPro, 2017a). In terms of products with multiple suppliers, GaBi seems to be a better option than SimaPro because it can construct a model with a range of different suppliers from its LCI datasets, while this is not possible for SimaPro.

Another noteworthy aspect is the connection between different products, which is currently unachievable in SimaPro because of its automatic creation of a product system (Rodríguez & Ciroth, 2016). Some features in GaBi help the

Table 2.5 Comparison of features between SimaPro and GaBi

Features		*SimaPro*	*GaBi*
LCI database	Eco-Invent	☒ (Ecoinvent, 2017; SimaPro, 2017b)	☒ (Ecoinvent, 2017; GaBi, 2017b)
	U.S.LCI	☒ (National Renewable Energy laboratory, 2012; SimaPro, 2017b)	☐ (National Renewable Energy laboratory, 2012; GaBi, 2017b)
	AusLCI	☒ (Life-cycle Strategies, 2017; SimaPro, 2017b)	Users need to contact Thinkstep (Australian Life-cycle Assessment Society (ALCAS), 2017; GaBi, 2017b)
Modelling	Products without provider	☐	☒
	Model loops	☒	☒
	Connections between different products	☐	☐
Ecological impact analysis methods	Midpoint approach methodologies		
	CML 37	☒	☒
	EDIP 2003 39	☒	☒
	TRACI 40	☒	☒
	Endpoint approach methodologies		
	Eco Indicator 99 41	☒	☒
	EPS 2000 43	☒	☒
	Eco Scarcity method (Eco points) 44	☒	☒
	JEPIX 45	☒	☒
	Combined midpoint and endpoint approach		
	RECIPE 46	☒	☒
	ILCD	☒	☒
	Impact 2002+ 49	☒	☒
	LIME 47	☒	☒
	LUCAS 51	☒	☒

Note: ☒ denotes that the feature is enabled on the platform

program by providing more optimized results than SimaPro, which could change the planner's decision in the earlier phase. These features include higher aquatic eutrophication and lower global warming impact (Jørgensen et al., 2013). However, Herrmann and Moltesen (2015) adjudicate that different results would be given by distinctive models developed by these platforms.

Although a variety of tools exist for developing building energy simulation models, researchers worldwide prefer to use the two leading and most commonly used life-cycle assessment programs, GaBi and SimaPro. These programs are preferred to other programs for the purpose of estimating the energy use in building operations due to their very large life-cycle database as well as their flexibility in designing many types of environmentally friendly products and systems (Herrmann & Moltesen, 2015; Židonienė & Kruopienė, 2015; Sinha et al., 2016; GaBi, 2017b; SimaPro, 2017b). Recent studies that employed these programs to assess life-cycle greenhouse gas emissions, as well as life-cycle energy consumption, are summarized in Table 2.6. A standard life-cycle assessment established

Table 2.6 Models developed for life-cycle environmental assessment

Author (year)	*Variable*	*Program*	*Building type*	*Objective*	
Sim et al. (2016)	Materials properties	GaBi	Residential	i	
Kua and Lu (2016)	Materials properties	GaBi	Residential	i	
Bueno et al. (2016)	Materials properties	GaBi	Multipurpose	i	
Lawania and Biswas (2016)	Building envelope	SimaPro	Residential	i	ii
De Souza et al. (2016)	Building envelope	SimaPro	Multipurpose		ii
Islam et al. (2016)	Building envelope	SimaPro	Multipurpose		ii
Wang et al. (2015)	Energy consumption	SimaPro	Residential	i	ii
Mangan and Oral (2015)	Energy consumption	GaBi	Retrofit	i	ii
Georges et al. (2015)	Building envelope	SimaPro	Residential	i	
Islam et al. (2015b)	Building envelope	SimaPro	Residential	i	ii
Weißenberger et al. (2014)	Energy consumption	GaBi	Zero energy buildings	i	ii
Desideri et al. (2014)	Energy consumption	SimaPro	Multipurpose	i	ii
Condeixa et al. (2014)	Building envelope	GaBi	Residential	i	
Asdrubali et al. (2013)	Energy consumption	SimaPro	Multipurpose	i	ii
Ximenes and Grant (2013)	Materials properties	SimaPro	Residential	i	
Basbagill et al. (2013)	Materials properties	SimaPro	Multipurpose	i	
Kulahcioglu et al. (2012)	Materials properties and energy consumption	GaBi	Multipurpose	i	ii
Iyer-Raniga and Wong (2012)	Energy consumption	SimaPro	Retrofit	i	
Gaidajis and Angelakoglou (2011)	Energy consumption	GaBi	Office		ii
Broun and Menzies (2011)	Materials properties	SimaPro	Multipurpose		ii
Blom et al. (2011)	Energy consumption	SimaPro	Residential	i	ii
Islam et al. (2010)	Materials properties	SimaPro, Accurate	Residential	i	ii
Ortiz-Rodríguez et al. (2010)	Materials properties	GaBi	Residential	i	ii
Dawood et al. (2009)	Building envelope	SimaPro	Multipurpose	i	ii
Techato et al. (2009)	Energy consumption	GaBi	Retrofit	i	ii

Note: (i): life-cycle greenhouse gas emissions; (ii): life-cycle energy consumption;

by the International Organization for Standardization (2006), which includes the goal and scope of the research, LCI, life-cycle impact assessment, and interpretation, is available (Al-Ghamdi & Bilec, 2016). These two life-cycle assessment tools are valid within the boundaries of the building life-cycle (Linfei et al., 2011; Zhang & Wang, 2015), mainly during the processes that consider the energy used across all stages of the lifespan of the building: material production, maintenance and substitution, and end-life phase of life-cycle (Al-Ghamdi & Bilec, 2016; Tam et al., 2018b). Both applications can analyze the material using scenarios or building envelope options of a real building to simulate its energy consumption as well as to generate estimate of greenhouse gas emissions during its lifetime (Islam et al., 2010; Broun & Menzies, 2011; Ximenes & Grant, 2013; Bueno et al., 2016; Islam et al., 2016; Kua & Lu, 2016; Sim et al., 2016). An analysis of external wall structure systems in the context of Brazilian buildings was conducted by Bueno et al. (2016) using GaBi and the Ecoinvent database 2.01. This life-cycle assessment study compared the results of the assessment of four structure types: masonry of clay blocks, masonry of concrete blocks, concrete wall, and cement panels with steel framing. This was performed by five different life-cycle impact assessment approaches. Beyond these approaches, the Impact World+ method, which is the most updated initiative, was evaluated as the most appropriate methodology. However, this book also indicates that there are still deficiencies in practical life-cycle assessment tools to justify most environmental data around the world. Kua and Lu (2016) contribute an environmental assessment of windows that use tempered glass and polycarbonate. In this model, data for all the procedures in the production phase were multiplied by the databases in Ecoinvent, GaBi (GaBi, 2017a), and ICE (Hammond & Jones, 2008) by a factor of the impact caused by 1 MJ of the Singaporean and Malaysian energy mixes to that of the European mix.

The case study of a 25-storey building's lifetime in South Korea has been assessed by Sim et al. (2016) using GaBi. Following the analysis, concrete and steel contribute significantly to environmental impact in terms of global warming potential, acidification potential, and eutrophication potential, while the cement and glass also are the primary source of the ozone depletion potential. SimaPro is likewise used to develop several models to assess greenhouse gas emissions, as well as energy consumption. Lawania and Biswas (2016) recommend the optimal option among 60 envelope options for a typical house in Perth, Australia, using a life-cycle assessment approach employing SimaPro with the AusLCI database (Australian Life-cycle Assessment Society (ALCAS), 2017).

With regard to the case of an existing building, researchers can study the optimal renovating option that has least environmental impacts (Techato et al., 2009; Iyer-Raniga & Wong, 2012; Mangan & Oral, 2015). The approach that is mentioned in Table 2.6 can be used to automatically assess the life-cycle impact of a building in terms of the energy consumption (input data) in each phase, as well as the greenhouse gas emissions (output data) of these stages, which can be applied in GaBi and SimaPro (Desideri et al., 2014; Weißenberger et al., 2014). Mangan and Oral (2015) used GaBi and ICE to access life-cycle energy and life-cycle carbon emission for the renovation options associated with improving an existing residential building's energy performance in Turkey.

The "net-zero emissions" building performance in the Norwegian context was investigated for energy consumption in the operation phase and embodied emissions in materials by SimaPro with the Ecoinvent version 2.2 (Georges et al., 2015). According to the study, the electricity supply for the performance of these net-zero emissions buildings is extensively reduced along with greenhouse gas emissions and embodied energy in materials and construction. GaBi and SimaPro are not dedicated architecture design software programs, but rather whole-life-cycle estimation programs with an enormous material database inventories for over 60 countries, including the Australian LCI library (Architecture & Design, 2016; GaBi, 2017b; International Building Performance Simulation Association (IBPSA), 2017; SimaPro, 2017b). Material quantities, which must be supplied in the development of the LCI, can be estimated by both of these programs.

2.7 Summary

Until recently, sustainable development was recognized as a principal solution to environmental issues, considering the emergence of environmental issues into economic and political decision-making. However, the widely known sustainability framework is referred to as the triple bottom line, which focuses on social and economic development, as well as environmental preservation measures. The social aspects of sustainable development always coincide with the definition of the goals of social sustainability and the contribution of these goals to social development (Nicola et al., 2011). On the other hand, sustainable economic development is the process that attempts to satisfy the society's needs, but in a manner that preserves and maintains natural resources and the ecosystem for future generations, as mentioned in the environmental sustainability dimension (Soubbotina, 2004).

This chapter also reviews the appropriate approach for a building's life-cycle assessment along with achieving Green Star requirements in reducing greenhouse gas emissions during the lifetime of a building. The review shows that there is a variety of programs that can estimate life-cycle energy consumption and life-cycle greenhouse gas emissions. From all the vast and continuously updated life-cycle assessment databases provided by software programs for many countries and regions, GaBi and SimaPro meet the requirements of the input-output approach methodology suggested in the Green Star rating tool. Thus, designers can develop their projects easily to achieve the requirements posed in this green building rating tool in Australia. On the basis of the literature, the book plans to expand its research further to develop a model that can access the environmental impacts produced during the life-cycle of a building in the Australian context. The literature identifies the research gap in developing a life-cycle environmental impact assessment considering all the critical criteria associated with the amount of greenhouse gas emissions during the lifetime of a green building.

In the process of environmental impact assessment, the relationship between construction work breakdown and life-cycle greenhouse gas emissions analysis is coherent and unquestionable. Only in the case when building work items perform analysis at the level of the unit process is it possible for the environmental impact analysis to evaluate for all types of materials and resources used in green building projects. Therefore, it is fundamental to consider the environmental impact during the whole life of a building by separately analyzing several elements of the building. Therefore, the life-cycle greenhouse gas emissions assessment of a green building is chosen as the significant analysis of this research.

In the context of life-cycle greenhouse gas assessment, the life-cycle assessment framework considers ecological effects of a typical building and a green building from cradle-to-grave. The optimal methodology of life-cycle assessment enables to categorize, characterize, and evaluate the environmental impact. A variety of life-cycle greenhouse gases assessment models exist that can identify the optimal design solution. Nevertheless, these models mostly focus on some specified criteria of green buildings, while it is necessary to assess all elements of the building. This chapter identified the gap in the research for development of a life-cycle greenhouse gas emissions assessment, which considers all fundamental criteria involving environmental impacts in a green building in Australia.

3 Classification of Green Star credits and model development for the reduction of greenhouse gas emissions

3.1 Introduction

This chapter comprises four main sections. The first two sections illustrate the process of Green Star credits classification, the role of Green Star rating tool in the process of life-cycle greenhouse gas emissions assessment, and the allocation of credit points in this rating tool. The next section demonstrates the credits in Green Star – Design & As Built associated with life-cycle greenhouse gas emissions reduction. These credits include three significant materials (timber, concrete, steel) that are generally used in construction projects, and the impact of the building's envelope (roof, wall, and floor) to reduce energy consumption and greenhouse gas emissions.

This research aims to develop a computer-assisted model to stimulate materials used for structures in commercial buildings in different geographical zones in Australia. The model is built to assess life-cycle greenhouse gas emissions and to achieve the required points for greenhouse gas emission reduction in Green Star – Design & As Built. The proposed model can then be used to eliminate difficulties, as green building design can be automated, user-friendly, and effective. Companies in the construction industry can readily use this model to implement their green building strategies without long-lasting design processes. The other purpose of the study is to analyze Green Star credits related to greenhouse gas emissions reduction, which leads to the ultimate objective of applying the life-cycle assessment methodology to a different building envelope system in the Australian construction industry. The credits in the Green Star rating tool, which consider the greenhouse gas emission reduction, are identified and selected for application in the model. In this stage, the data of embodied greenhouse gas emissions are collected. Variables of the model are determined taking into account Green Star credits as well as the characteristics of commercial buildings to ensure the validity and reliability of results. After the model is developed along with the inventory analysis, the life-cycle greenhouse gas emissions results are evaluated and validated.

3.2 Credits classification methodology

This section presents a comprehensive study on Green Star rating tool's credits and their criteria points in order to analyze the credits involved in greenhouse gas

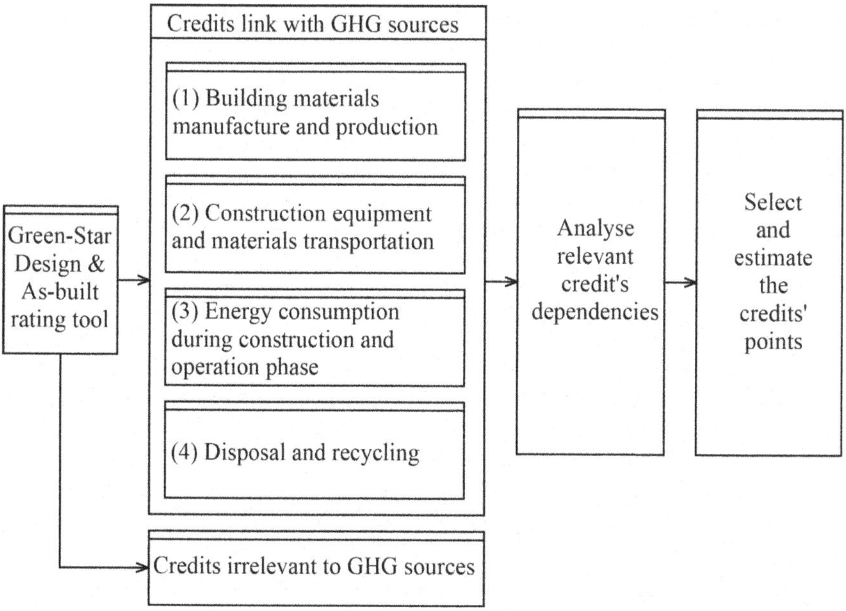

Figure 3.1 Flowchart of the credits classification process

emission reduction for a building during its lifetime and how many credit points can be achieved. Possible sources of released greenhouse gases in a building are: (1) building materials manufacture and production; (2) construction equipment and materials transportation; (3) energy consumption during the construction and operation phase; and (4) disposal and recycling. Therefore, all credits involved with these sources are assessed. The process of credits classification is illustrated in Figure 3.1.

After the credits related to greenhouse gas emissions are analyzed, the relationship between the credit's dependencies is also assessed. Finally, available points from these credits are awarded when the credits' pathway criteria are achieved.

3.3 Green Star design rating tool in Australia

3.3.1 *Role of Green Star in life-cycle greenhouse gas emissions assessment*

In Australia, the energy usage for heating and cooling accounts for about 18% of the average energy usage of an occupied office. Greenhouse gas emissions produced in a commercial buildings' lifetime are projected to increase along with the economic growth by mid-2020s (Australia Government, 2016). Legislative frameworks in Australia have been developed along with the effectiveness and prospects of green building rating tools. Most of these tools are voluntarily used, except in the case of some official regulations supporting rating systems such as the Building Sustainability Index (BASIX) and Nationwide House Energy Rating Scheme (NatHERS). BASIX is compulsory for projects in New South Wales regions, and NatHERS provides regulations for thermal efficiency. Further, NatHERS and BASIX ratings can be employed to achieve specific credit points in the Green Star Environmental Rating System (Green Star).

Australia was chosen for the analyses of this book because the country is ranked (1) eighth in the top ten greenhouse gas-emitting countries and second in the top ten countries with the most greenhouse gas emissions releasing per capita (528,106.28 tons of CO2 equivalents, thousands), and (2) third for total greenhouse gas emissions per 1,000 dollar GDP (0.576 kg CO2 equivalents per 1,000 USD, thousands) (Organisation for Economic Co-operation and Development (OECD), 2017).

It is indisputable that the green building implementation, which has been encouraged by the Green Building Council of Australia by means of the Green Star Environmental Rating System, can help improve the company image and ability to combat environmental destruction activities. Nevertheless, there are common thoughts among investors that the implementation of environment-friendly methods to achieve specific green building status will significantly alter their project's budget (Kats, 2003). They will only achieve minimal points required for a particular green building status. Therefore, the green building design should be widely encouraged and reinforced in the industry (Australian Inventory data project, 2009).

For the Australian sustainable development approach in general, and particularly for Green Star to implement life-cycle assessment of the buildings, construction projects should be assessed consistently with the requirements of the Building Energy Analysis Software protocol in the Australian Building Code Board (Australian Building Codes Board, 2006). The recent environmental analysis programs have their advantages and disadvantages in demonstrating every corner of the efficient building to comply with the protocol's requirements. Moreover, choosing which software to use in the early and middle phases of project design for evaluation of the energy efficiency and ecological impact is also a crucial issue to designers and project planners.

Computational models represent one of the current methods for addressing global environmental tasks by estimating life-cycle energy consumption and life-cycle greenhouse gas emissions. Although several building models for greenhouse gas emissions produced during a building's life have been proposed, the results generated using different environmental impact approaches remain incoherent and inconsistent (Herrmann & Moltesen, 2015). This implies that it is difficult to choose optimal design scenarios simultaneously for both environmental improvement and financial plans. This matter is also valid in the Australian context, where a plan of reducing about 26% of greenhouse gas emissions from 2005 to 2030 has been initiated (Shahiduzzaman & Layton, 2015).

3.3.2 *Credit points allocation in Green Star rating tool*

The Green Star Environmental Rating System was designed in 2003 (Green Building Council of Australia, 2005) with the aim of (1) establishing a standard benchmark and a standard of measurement for green buildings; (2) promoting integrated building design; (3) identifying building life-cycle impact; (4) raising awareness of green building benefits; (5) recognizing environmental leadership; and (6) transforming the construction industry to reduce its impact on the environment.

The system includes four green building rating tools, namely: "Green Star – Design & As Built", "the Green Star – Interiors", "the Green Star – Communities", and "the Green Star – Performance". As illustrated in Table 3.1, to be granted a Green Star certification, projects should achieve a rating of 4- to 6-star, which has a score from 45 to more than 75 points. With the lower overall score, only when using "the Green Star – Performance" rating tool, projects with 1-star, 2-star, and 3-star ratings are still awarded certification that the other rating tools do not allow.

Green Star – Design & As Built rating tool version 1.1 divides with hierarchical structure into nine major criteria: (1) management; (2) indoor environment quality; (3) energy; (4) transport; (5) water; (6) materials; (7) land use and ecology; (8) emissions; and (9) innovation. These criteria include 100 basis points, which are allocated in 30 credits listed in Table 3.2. These points are distributed when the assessed building meets the credits' requirements.

In the "Management" category, a project is rewarded by facilitating strategies as well as promoting a practical plan to achieve a project's sustainable potential. There are 14 credit points available in this category, where general information about the project is given as an adaptation of the project to the Green Star principles, targets, and methods to enhance the building's environmental performance during its life-cycle. The primary purpose of this category is to help and reward the implementation of practices and procedures that allow producing optimal practical sustainability results throughout the different phases of project design, construction, and its ongoing operation (Green Building Council of Australia, 2015a).

However, the management activities in Green Star involve mostly providing building information, environmental performance targets, and plans, which do not produce emissions directly during the life-cycle of buildings. In the criterion "Operational waste" under the "Management" category, there are two pathways to achieve credit points. The pathway "8B Prescriptive Pathway"

Table 3.1 Green Star rating score

Rating	*Overall score*	*Result*	*Remarks*
0-star	<10	Assessed	Not granted certification
1-star	10–19	Minimum practice	
2-star	20–29	Average practice	
3-star	30–44	Good practice	
4-star	45–59	Best practice	Only projects with a higher than 4-star rating are awarded certification
5-star	60–74	Australian excellence	
6-star	>75	World leadership	

Table 3.2 Green Star categories and allocation of credit points

Item	*Category*	*Criterion*		*Credit points*
I	Management	1	The Green Star accredited professional	1
		2	Commissioning and tuning	4
		3	Adaptation and resilience	2
		4	Building information	2
		5	Commitment to performance	2
		6	Metering and monitoring	1
		7	Construction environmental management	1
		8	Operational waste	1
II	Indoor environment quality	9	Indoor air quality	4
		10	Acoustic comfort	3
		11	Lighting comfort	3
		12	Visual comfort	3
		13	Indoor pollutants	2
		14	Thermal comfort	2
III	Energy	15	Greenhouse gas emissions	20
		16	Peak electricity demand reduction	2
IV	Transport	17	Sustainable transport	10
V	Water	18	Potable water	12
	Materials	19	Life-cycle impacts	7
VI		20	Responsible building materials	3
		21	Sustainable products	3
VII	Land use and ecology	22	Construction and demolition waste	1
		23	Ecological value	3
		24	Sustainable sites	2
		25	Heat island effect	1
VIII	Emissions	26	Stormwater	2
		27	Light pollution	1
		28	Microbial control	1
		29	Refrigerant impacts	1
IX	Innovation	30	Innovation	10

requires the assessed project has used "best practice assessment" facilities to collect waste. This pathway has one point, similar to the pathway "8A Performance Pathway", which is used as an operational waste management plan in the building's design. Thus, although the pathway "8B Prescriptive Pathway" is not used in the research, one credit point remains, with one point available. Therefore, the credits in the "Management" category are excluded from the research.

As illustrated in Figure 3.2, 22% of credit points are allocated to the "Energy" category, 14% belong to "Materials", and 17% belong to the "Indoor environment" criteria (Green Building Council of Australia, 2015a; Tam et al., 2017b). These categories represent a significant proportion of credit allocation under

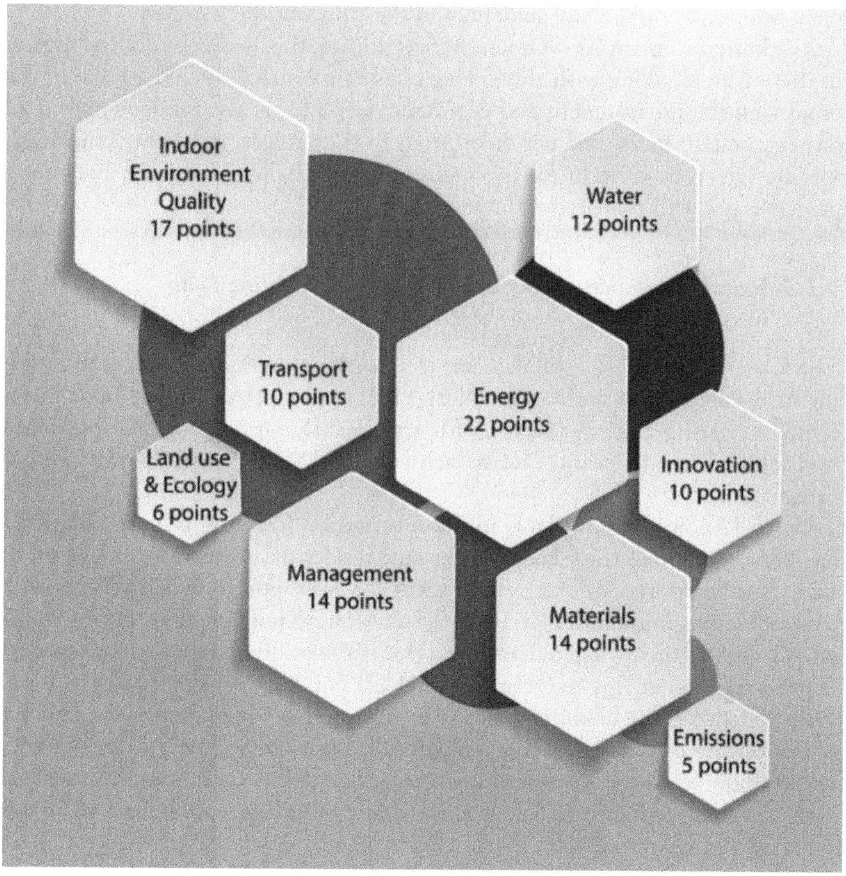

Figure 3.2 Credit point allocation in Green Star

Green Star. These multi-objective elements, which relate to energy-linked greenhouse gas emissions reduction during a buildings life-cycle, are correlated with the requirement of the criteria under the Green Star – Design & As Built rating tool (Green Building Council of Australia, 2015a). Six criteria with 17 credit points in the "Indoor environment quality" category focus on matters such as quality of indoor air, lighting, and thermal and acoustic comfort. "Transport", "Water", "Land use and ecology", and "Emissions" categories occupy 10, 12, 6, and 5 points, respectively.

The extra 10 points in the "Innovation" category are additional points because they are not included in the list of common points and can be achieved by utilizing strategies or initiatives following this category's guidance. The additional points can only be achieved after common credit points in the other eight categories are achieved, and the project is verified to have an innovative design, plan, or initiative following guidelines in the "Innovation" category.

In addition to granting a Green Star certificate, the assessment of the impact of these factors, along with the energy consumption throughout the life of the project on the environment and economic decisions, is also fundamental to all owners, practitioners, and persons related to the project. As such, achieving a specific Green Star credit for the industry seems challenging if the assessment process is not well-prepared and user-friendly.

3.3.3 *Relationship between Green Star and other rating tools in Australia*

The Green Star rating tool also connects with other Australian green building assessment tools such as the NatHERS, National Australian Built Environment Rating System (NABERS), and BASIX rating tool. As illustrated in Table 3.3, credit points can be achieved in "Management" and "Energy" categories.

Up to 12 points are available in the subcriterion Credit 15B NatHERS. This method only applies to Class 2 multi-unit residential dwellings located in all states and territories in Australia, except for New South Wales. The project must achieve a minimum NatHERS 0.5-Star rating improvement on the minimum legislated area-weighted average. This pathway divides into two categories: (1) thermal and energy performance; and (2) building services, sealing testing, and appliances. The credit element of energy intensity reduction is awarded six points on the calculated energy reduction intensity in the base MJ/m^2 metric, rather than the NatHERS star rating score. Nine points of the specification and design checks for lighting, domestic hot water, building sealing, and appliance

Table 3.3 Link between Green Star and other assessment tools

Item	*Category*	*Criterion*	*Subcriterion*	*Credit points*	*Linked tool*
I	Management	5. Commitment to performance	5.1 Environmental building performance	1 point	NABERS Energy Commitment Agreement
III	Energy	15. Greenhouse gas emissions	15B GHG emissions reduction – NatHERS	12/20 points	NatHERS
			15C GHG emissions reduction – BASIX	16/20 points	BASIX
			15D GHG emissions reduction – NABERS Energy Commitment Agreement	16/20 points	NABERS Energy Commitment Agreement

Table 3.4 Points awarded under Credit 15, section B

Credit element	*Greenhouse gas emissions reduction (%)*	*Awarded points*
Greenhouse gas emissions in comparison with BASIX GHG emissions benchmark	20%	3
	40%	6
	60%	10
	80%	13
	100%	16

efficiencies, as well as ventilation systems, are considered; however, a maximum of six points may be awarded.

Unlike the 15B pathway, Credit 15C – Greenhouse gas emission reduction – BASIX is used for projects classified as National Construction Code of Australia's class 1 and 2 residential houses. When the buildings thermal performance is better than the allowed BASIX greenhouse gas rate, credit points can be calculated by Green Star: Greenhouse gas emissions calculator (Table 3.4). The Credit 15C BASIX also applies to Class 2 multi-unit residential dwellings but only for projects located in New South Wales. The method can be used for Class 1 residential houses. Nevertheless, the relevant clauses need to be adapted to the relevant BASIX compliance criteria applicable to these building types and their construction methods. Greenhouse gas emission reduction of the project through thermal performance improvement and efficient building services, as well as appliances and equipment utilization, can be awarded up to 16 points in this pathway. Points are calculated using the Green Star performance of greenhouse gas emissions.

As with Credit 15C, Credit 15D NABERS can likewise achieve 16 points. Regarding Credit 15D, points are achieved on the estimated reduction in the NABERS baseline and interpolated using the Green Star (Design & As Built) greenhouse gas emissions calculator (Green Building Council of Australia, 2015a, 2015b).

3.3.4 *Additional pathways in Green Star rating tool*

Table 3.5 shows certain credits that contain additional points whenever the condition of this criterion is achieved. For example, in the pathway criterion 19A.2 "Additional life-cycle impact reporting" under the "Materials" category, there is one point available to be awarded to the project if this project provides five additional life-cycle impact assessments, including Human toxicity, Land use, Resource depletion, Water stress indicator, Ionizing radiation, and Particulate matter. These environmental impacts are essential factors in the life-cycle environmental assessment to help designers obtain a better perspective to achieve green building design (Green Building Council of Australia, 2014b; LCEE life-cycle Engineering Experts GmbH, 2016).

Table 3.5 Green Star additional credits

Item	*Category*	*Criterion*	*Subcriterion*	*Credit points*	*Condition for award*
I	Management	2. Commissioning and tuning	2.4 Independent commissioning agent	1 point	One of the credits 2.1, 2.2 or 2.3's requirements meet
II	Indoor environment quality	12. Visual comfort	12.1 Daylight	1 point	80% of the nominated area demonstrate high daylight levels
		13. Indoor pollutants	13.1. Paints, adhesives, sealants, and carpets	1 point	More than 50% of paints have an average of 5 g/L TVOC content
		14. Thermal comfort	14.2. Advanced thermal comfort	1 point	95% of the designated area of thermal comfort is given
VI	Materials	19. Life-cycle impacts	19A.2 Additional life-cycle impact reporting	1 point	Reporting five additional impact factors
VIII	Emissions	26. Stormwater	26.2. Reduced pollution targets	1 point	When the 1st point is achieved, and rainwater complete discharge of rainwater from project site

3.4 Life-cycle greenhouse gas emissions assessment methodology

3.4.1 *Green Star credits related to life-cycle greenhouse gas emissions reduction*

Climate change requires the immediate implementation of practical actions (Müller & Harnisch, 2008). One of the major fundamental issues for the building and construction industry is to address global climate change challenges by developing credible greenhouse gas estimation schemes for building materials (Wu et al., 2014; Cho & Chae, 2016). Moreover, concrete is the most widely used construction material (1) whose demand cannot be met by the world's cement production, and (2) which currently contributes to about 5% of the annual global anthropogenic carbon dioxide emissions (Flower & Sanjayan, 2007; Kline & Kline, 2015).

Nowadays, the construction industry tends to move toward sustainable building design, which is defined as "the ability to fulfil the needs of the future" (American Society of Heating Refrigerating and Air-Conditioning Engineers Press, 2006) and covers the entire building life-cycle from raw material extraction to design, construction, operation, and demolition (Zimmermann et al., 2005; Wang et al., 2011). In this regard, the promotion of green strategies in a building's life-cycle plays a significant role in achieving sustainability (Zhang et al., 2011).

A sustainable construction industry should be marked by minimal environmental impact not merely during the manufacturing processes, but also during the entire life-cycle of the projects (Sabnis, 2015). To promote design and construction practices that reduce adverse environmental impacts on buildings and improve the occupants' health and well-being, many green building rating systems have been developed, including: (1) the Leadership in Energy and Environmental Design (LEED) by the United States Green Building Council (USGBC) (Farham & Gholian, 2014); (2) Building Research Establishment Environmental Assessment Method (BREEAM) by the Building Research Establishment (O'Malley et al., 2014); (3) the Comprehensive Assessment System for Building Environmental Efficiency (CASBEE) by the Japan Sustainable Building Consortium (Wong & Abe, 2014); and (4) the Green Star Environmental Rating System by the Green Building Council of Australia (Sabnis, 2015).

Life-cycle greenhouse gas emissions assessment of building material, including concrete, follows three internationally recognized standard systems: PAS 2050:2011 (British Standards Institution, 2011), ISO 14040:2006 (International Organization for Standardization, 2006), and a greenhouse gas protocol (Flower & Sanjayan, 2007; Greenhouse Gas Protocol, 2011; Chau et al., 2015; The Concrete Network, 2016). The environmental impact of construction material occurs at many stages in production: construction, use, and the end-of-life-cycle stage of material. Life-cycle assessment may be a lengthy calculation process, which can be tedious and deter companies, industry partners, and designers in implementing the green building design. As such, achieving the specific Green Star credit for the industry seems difficult because the process has not been well-prepared and user-friendly. The proposed model has therefore been developed and pursued with this difficulty in mind, so that green building design is ultimately automated, user-friendly, and effective. As a consequence, companies can readily use it to implement their green building strategies without lengthy design processes.

Examination of life-cycle greenhouse gas emissions can be divided into three stages following the entire building's life-cycle from the perspective of material and energy flow: (1) the materialization stage: incorporating material preparation, transportation, and on-site production; (2) the operational stage: daily usage, daily routine, and engineering renovation; and (3) the disposal stage: incorporating building demolition, waste transportation, and material recycling (Zhang & Wang, 2015).

Life-cycle assessment should be carried out through the entire lifetime of the building from product manufacture, construction, operation, destruction, and disposal, till the recycling stage of the project. The input data, as recommended in Chapter 2, include the raw materials used and the energy consumption of the projects. The environmental impact of materials mainly stems from concrete, steel, and timber (Junnila & Horvath, 2003; Asif et al., 2007; Zabalza Bribián et al., 2011). The environmental contribution of these major materials is approximately 30%, 20%, and 2% for concrete, steel, and timber, respectively, for 1 m^2 of construction area (Zabalza Bribián et al., 2011). Credit 13 – Indoor pollutant, 19 – Life-cycle impacts, and 20 – Responsible building material in Green Star also include the impact of materials in order to ensure that the buildings have a sustainable design. These credits contribute 17 points to the total points of a green building.

Total building energy consumption in Australia is projected to rise by 24% over the period 2009–2020, with the expected total energy output just under 170 petajoules (PJ) by 2020, as reported in the National Strategy on Energy Efficiency by the Council of Australian Governments (2012). The Green Star Environmental Rating System is one of the green building rating systems in Australia. The "Energy" criterion occupies 24 out of 100 points in the Green Star (Design & As Built). Among these points, to encourage the reduction of greenhouse gas emission associated with energy usage in building operations, up to 20 points can be achieved in Credit 15 alone.

The Green Star rating tool, governed by the Green Building Council of Australia (GBCA), is designed to advocate the construction field in its sustainable and eco-friendly development. Green Star includes eight main categories for evaluating green buildings. Table 3.6 illustrates the credits related to greenhouse gas emission analysis during the life of buildings.

The primary purpose of the "Indoor environment quality" category is to help the projects achieve improvements in occupants' well-being and sustainability performance. Credit 13 – Indoor pollutants and 20 – Responsible building material (timber product assessment), which are related to timber use in the building, may gain two available points when the project utilizes at least 95% of reused wood products or no new wood material (Green Building Council of Australia, 2015a).

Credit 15 with the greenhouse gas emission reduction performance under the Green Star category "Energy" is designed to facilitate efficient energy consumption

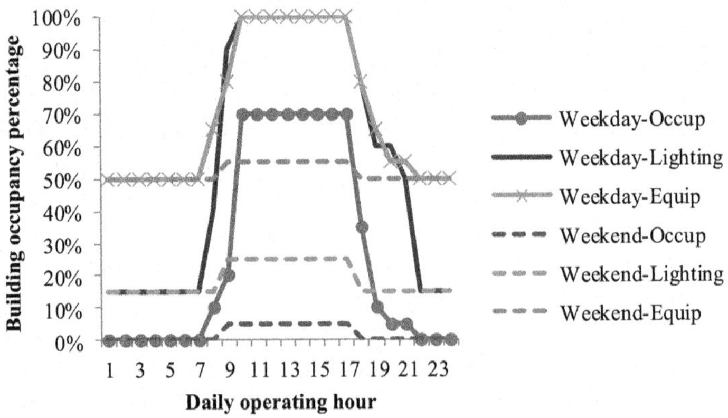

Figure 3.3 Occupancy trends in office operating schedules

Table 3.6 Green Star credit points regarding greenhouse gas emissions analysis

Category	*Credits*	*Credit point(s)*	*Involved aspects in LCA*
Indoor environment quality (IEQ)	13 – Indoor pollutants	1	Timber products
Energy	15 – Greenhouse gas emissions	20	Building envelops and energy consumption
Materials	19 – Life-cycle impacts	12	Concrete, reinforcement bar used in buildings
	20 – Responsible building material	1	Timber products
	Total possible achieved credit points	34	

to reduce greenhouse gas emissions released during the operational stage and buildings' life-cycle, which accounts for 24 out of 100 points under Green Star. Twenty points will be granted to Green Star projects that utilize energy performance strategies, and extra four points will be achieved using the buildings' peak demand load reduction on its electrical systems to reduce greenhouse gas emissions (Green Building Council of Australia, 2015a). The 100-year global warming potential (GWP100) is used to describe global warming potential over the period of 100 years for the purpose of determining the Green Star rating tools (Green Building Council of Australia, 2014a).

The pathway "Modelled performance" in this credit is applied to buildings of Class 2 to 9 in Australia. In this pathway, the proposed building greenhouse gas emissions should be demonstrated to be less than those of the equivalent benchmark building. Up to 20 points can be achieved in this pathway, and 4 out of these points can be granted from improving the building's fabric against a reference building. Further 16 out of 20 points can be granted for reducing emissions against the benchmark building. In this book, the components of the building system, which are complied with the requirements of the handbook issued by the Insulation Council of Australia and New Zealand (2014), as well as the regulations of the Building Code of Australia, AS/NZS 4859.1, will be analyzed. These components include indoor and outdoor air films, building materials used in the system, and air spaces. Up to 20 points can be obtained, 16 points for reducing greenhouse gas emissions against the benchmark building, and the other 4 points from developing the building envelope against a reference building. A guide to the credit points that can be awarded in Credit 15 is summarized in Table 3.7.

Green Star also aims to use life-cycle assessment approaches to encourage practitioners and designers demonstrating their projects' ability to perform sustainability in the construction project life-cycle (Green Building Council of Australia, 2015c; Tam et al., 2017c). A "Life-cycle impacts" criterion is thus divided into two parts: Life-cycle assessment and Life-cycle impacts of materials.

Table 3.7 Guidelines to achieve points under Green Star's Credit 15

Credit element	*Reduction (%)*	*Points awarded*
Energy consumption reduction (intermediate building relative to reference building)	5%	1.0
	10%	2.0
	15%	3.0
	20% (maximum rewarded)	4.0
Greenhouse gas emissions reduction (proposed building relative to benchmark building)	10%	1.6
	20%	3.2
	30%	4.8
	40%	6.4
	50%	8.0
	60%	9.6
	70%	11.2
	80%	12.8
	90%	14.4
	100% (maximum rewarded)	16

The "Life-cycle assessment" can obtain six points when projects are implementing life-cycle assessment with a "whole-of-building" and "whole-of-life" methodology in compliance with the National Construction Code of Australia (Green Building Council of Australia, 2014b; Australian Building Codes Board, 2016; Le et al., 2018a). This credit involves six major impact factors, including stratospheric ozone depletion potential, acidification potential of land and water, eutrophication potential, tropospheric ozone formation potential, photochemical ozone creation potential (POCP ethylene equivalents), and mineral and fossil fuel depletion (abiotic depletion), and five additional impact categories: human toxicity, land use, resource depletion – water, ionizing radiation, and particulate matters. This means that the credits allow the project to claim two points under the "Project Life-cycle Impact Assessment", and the "Additional Life-cycle Impact Reporting".

The "Life-cycle assessment" has six points available, of which three points would be achieved when planners apply methods to reduce the environmental impacts of concrete. The credit is employed to recognize and encourage the reduction of greenhouse gas emissions, resource usage, and reduce wastage associated with concrete usage (Green Building Council of Australia, 2015a). The credit addresses all types of concrete used in a project, including structural and non-structural elements except concrete masonry. Up to two credit points can be achieved where the Portland cement content in the concrete used in the project has been reduced by replacing it with supplementary cementitious materials (SCM). Fly ash, blast furnace slag, and silica fume are selected as SCM as recommended by the Green Building Council of Australia.

Extra credit points can further be awarded if Portland cement content is reduced by 30–40%. This content is measured by the total mass of all concrete used in the project compared to the reference case as specified in the Green

Building Council of Australia's technical manual (which has a range 280–610 kg for 1 m^3 concrete, for concrete which has a strength 20–100 MPa, respectively) (Green Building Council of Australia, 2015a).

An extra half a credit point is available if the water mix for all concrete used in the project contains at least 50% of captured or reclaimed water (measured across all concrete mixes in the project). The remaining half a point can be obtained if the following criteria are met: (1) at least 40% of coarse aggregate in the concrete is crushed slag aggregate or another alternative material (measured by mass across all concrete mixes in the project), provided that the use of such material does not increase the use of Portland cement by over 5 kg/m^3 of concrete; and (2) at least 25% of fine aggregate (sand) inputs in the concrete are manufactured sand or other alternative materials (measured by mass across all concrete mixes in the project), provided that the use of such material does not increase the use of Portland cement by over 5 kg/m^3 of concrete. In Australia, both coarse and fine aggregates are materials derived from the processing of recycled concrete from construction and demolition waste (Cement Concrete & Aggregates Australia, 2008). Up to two points can be achieved from pathway 19 – Life-cycle impacts – Steel, when there is a reduction in the mass of steel reinforcement used compared to standard practice. The project's design team demonstrates in their quantity breakdown report the deduction of steel quantity with the standard level to achieve the credit points.

3.4.2 *Model development for life-cycle greenhouse gas emissions along with Green Star selected criteria*

In the Green Star – Design & As Built rating tool, there are eight available main categories. From these categories, "Indoor Environment", "Energy", and "Materials" criteria contribute 17%, 22%, and 14%, respectively, of the total 100 credit points (Green Building Council of Australia, 2015a). This represents a significant proportion of credit allocation. Therefore, to obtain a certificate under the Green Star Environmental Rating System, it appears essential to focus on the credits in these categories. For these credit points, Green Star focuses on materials itself, such as concrete, steel, and timber, and furthermore discusses the lifetime of a building in terms of production, construction, demolition, and recycling stages.

The model development flow in Figure 3.4 illustrates how the quantity of the project is input into the calculation process of greenhouse gas emissions and how the relevant credit points concerning greenhouse gases can be achieved. The process of the model development contains the following:

(1) The location of the project is defined to determine the difference between outdoor and indoor temperature, which is useful for estimating the amount of energy consumption for office heating and cooling.
(2) Based on the project design, work items are analyzed to define the quantity of main materials that are used in this project.

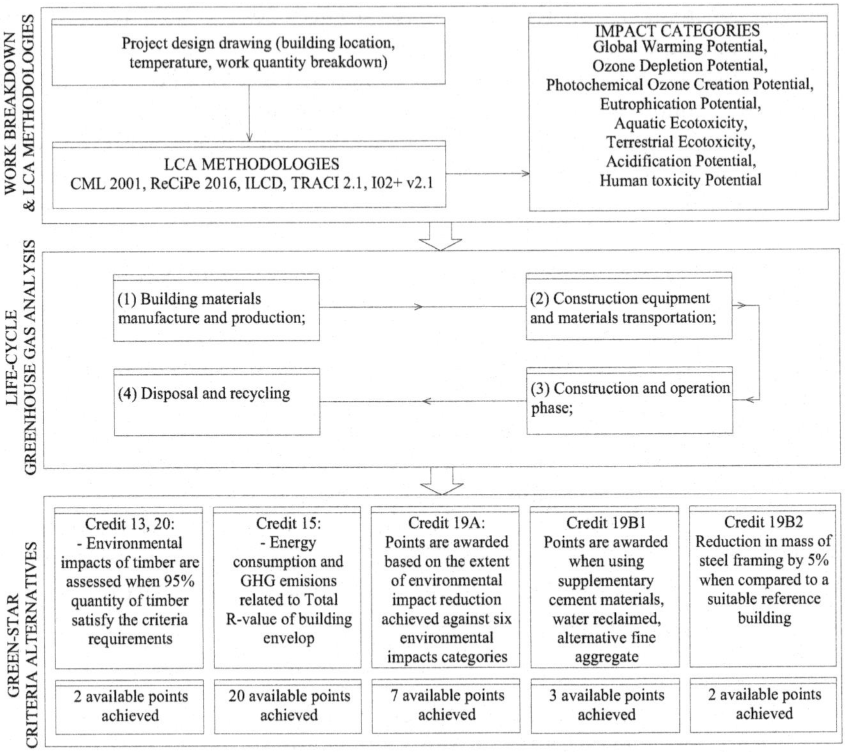

Figure 3.4 Model development flow

(3) The life-cycle greenhouse gas emissions analysis is implemented for the item unit (i.e., 1 m^3 timber, 1 m^3 concrete, or 1 kg steel, etc.), which includes the processes of production, construction, operation and maintenance, demolition, and recycling. The results are calculated per the Equation 3.1 based on the unit's life-cycle greenhouse gas emissions by the material's quantity, which is estimated from the provided design.

$$LCG = GHG_{unit} \times Q_{material} \quad \text{Equation 3.1}$$

where LCG denotes the amount of life-cycle greenhouse gas emissions, GHG unit denotes the amount of greenhouse gas emissions of 1 unit of each material, and Qmaterial denotes the quantity of material.

(4) At the final step, Green Star credit points are calculated by applying the requirements of each credit.

The quantity of life-cycle greenhouse gas emissions from some buildings, other than commercial office buildings, is calculated to determine the lowest level of greenhouse gas emissions and the associated environmental impacts of the buildings' materials (Equation 3.2). Variations of the model relate to the materials used in the building, the elements of the building's envelope, building location, and outdoor temperature.

$$\sum GHG = GHG_{timber} + GHG_{steel} + GHG_{concrete} + GHG_{energy\_consumption}$$

$$\sum Credit\_point = P_{C13} + P_{C15} + P_{C19} + P_{C20} \qquad \text{Equation 3.2}$$

The selection of impact categories of the book complies with the guidance in the credit: life-cycle impact in the Green Star – Design & As Built rating tool, including global warming potential, ozone depletion potential, photochemical ozone creation potential, eutrophication potential, aquatic ecotoxicity, terrestrial ecotoxicity, acidification potential, and human toxicity potential (Green Building Council of Australia, 2015a).

To assess the construction project's environmental impact, and use alternative solutions, designers require software that integrates and harmonizes life-cycle assessment methods and databases to develop their projects in order to reduce uncertainties of energy consumption and greenhouse gas emissions analyses. GaBi, developed by Thinkstep with the updated version 8.7, which is one of the flagship life-cycle assessment programs, has flexible environmental databases that can effectively satisfy the intended model's requirements (Tam et al., 2018a).

With coherent and logical libraries on environmental impact analyses, such as CML 2001, ReCiPe 2016, Tool for the Reduction and Assessment of Chemical and other environmental Impacts (TRACI), Impact Assessment of Chemical Toxic (IMPACT) 2002+, and the International Reference Life-cycle Data System (ILCD), it is possible for designers to incorporate these available resources within the life-cycle assessment frameworks. These methodologies provide information on environmental impact factors at midpoint and endpoint levels, which are useful for the proposed analyses. Life-cycle methodologies available for use in GaBi include CML 2001, ReCiPe 2016, ILCD, TRACI 2.1, and I02+ v2.1 (Table 3.8) (Zabalza Bribián et al., 2009; Owsianiak et al., 2014; Tam et al., 2018a).

The book's motivation was to help designers using other tools to assess their projects for reduced adverse environmental impacts by analyzing the life-cycle of buildings' envelopes. This model is designed to customize, following the Green Star rating tool, and assist project's stakeholders in finding a way to achieve this system certification for their projects. The intended model is designed to develop a library of frequently used material for building envelopes in Australia, as well as the process to estimate greenhouse gas emissions in the building's life-cycle in GaBi. In this manner, designers can drag and drop the option they want to choose for their project from the library to achieve all results related to the needed to mitigate environmental impact. The resources that are assessed in the model are explained below.

3.4.2.1 *Timber products and relevant Green Star credits – Credits 13 and 20*

Timber is a common material used in construction, although in the modern building this material is gradually being replaced by other materials such as

Table 3.8 The impact categories selected for the study

Impact category	*Category unit*	*LCA method*
Global warming potential	kg CO2-eq.	ReCiPe 2016, ILCD, TRACI 2.1, I02+ v2.1
Ozone depletion potential	kg CFC 11-eq	ReCiPe 2016, ILCD, TRACI 2.1,
	kg R-11 eq.	I02+ v2.1
Photochemical ozone creation potential	kg NOx eq.	ReCiPe 2016
	kg NMVOC eq.	ILCD
	kg C2H4 eq. to air	I02+ v2.1
Eutrophication potential	kg PO4-eq	ReCiPe 2016, ILCD, I02+ v2.1
	kg N eq.	TRACI 2.1
Aquatic ecotoxicity	kg 1,4 DB eq.	ReCiPe 2016
	CTUe	ILCD, TRACI 2.1
	kg TEG eq. to water	I02+ v2.1
Terrestrial ecotoxicity	kg 1,4-DB eq.	ReCiPe 2016
	kg TEG eq. to soil	I02+ v2.1
Acidification potential	kg SO2-eq	ReCiPe 2016, TRACI 2.1, I02+ v2.1
	Mole of H^+ eq.	ILCD
Human toxicity potential	kg 1,4 DB-eq	ReCiPe 2016
	CTUh	ILCD, TRACI 2.1

concrete and steel (Ramage et al., 2017). The life-cycle inventory (LCI) assessed for timber products in this book embodies the quantitative outcomes for several major timber materials used in Australia (Puettmann & Wilson, 2007; Robertson et al., 2012). The environmental assessments are energy consumption and environmental impacts for the extraction, manufacture, and transportation of timber resources.

A variety of studies in Australia and New Zealand show that the life-cycle greenhouse gas impact of timber products are considerably lower than other materials used in a construction project (Buchanan & Ievine, 1999; Carre, 2011; May et al., 2012). The timber products being used in Australia are mainly disposed to landfills at the end of their service lifetime. The quantity of waste from wood products is forecast to increase progressively to about 2 Mt by 2050 (Mclennan Magasanik Associates, 2010).

The life-cycle greenhouse gas emissions assessment is conducted using GaBi 8.7 with the functional unit for the analysis of 1 m^3 of timber in 50 years of the building's life (Figure 3.6).

Equation 3.3 shows the calculation of the impact category indicator, which is the sum of the products of inventory data with their characterization factors.

$$I_c = \sum \text{Inventory\_data} \times \text{CF} \qquad \text{Equation 3.3}$$

where I_c denotes impact category indicator and CF are the characterization factors that present a single emission's potential to contribute to a specific impact category (International Organization for Standardization, 2006). The characterization factors applied to timber products are summarized in Table 3.9.

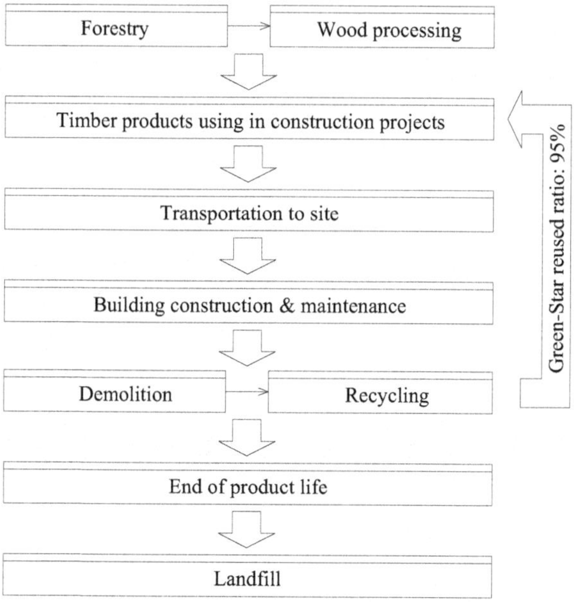

Figure 3.5 Flowchart of wood product system boundaries

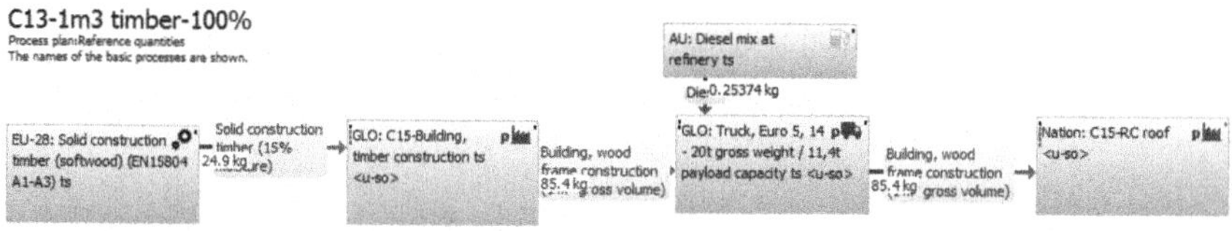

Figure 3.6 Analysis for 1 m^3 timber – (the model's module flow in GaBi 8.7)

The life-cycle impact results are analyzed with the inventory data from the Australian Life Cycle Inventory library and the life-cycle assessment methods of CML2001, ILCD, ReCiPe 2016, and TRACI 2.1 in GaBi 8.7. After the analysis, designers can use Equation 3.3 to estimate the environmental impact from the quantity of timber, which can be acquired from the building quantity breakdown.

3.4.2.2 Energy consumption and GHG emissions reduction – Credits 15 and 19

Typical envelopes for commercial buildings in combination inside Australian climate zones are deployed in the proposed estimation model to assess life-cycle greenhouse gas emissions (Australian Building Codes Board, 2016; Ga.gov.au, 2016). Climate temperatures directly affect buildings heating and cooling, and thus affect greenhouse gas emissions. These life-cycle-assessment categories are in

Table 3.9 Impact categories and LCA classification and characterization factors of GHGs

Impact category	*LCA classification*	*Characterization factor*	*Reference*
Global warming potential (GWP 100 years)	Carbon dioxide (CO2)	1 kg = 1 kg CO2 eq	(Heijungs et al., 1992; Houghton et al., 1996)
	Methane (CH4)	1 kg = 21 kg CO2 eq	
	Dinitrogen oxide (N2O)	1 kg = 264 kg CO2 eq	
Aquatic ecotoxicity	Ethene (C2H4)	1 kg = 4.88×10^4 kg TEG eq. to water	(Owsianiak et al., 2014)
Eutrophication potential (EP)	Phosphorus (PO4)	1 kg = 1.3 PO4 eq	(Heijungs et al., 1992; Zelm, 2009; Bare, 2012)
	Nitrogen oxides (NOx)	1 kg = 1 kg N eq	
Human toxicity	Sulphur dioxide (SO2), 1,4 dichlorobenzene (C6H4Cl2)	1 kg = 0.096 C6H4Cl2 eq	(Acero et al., 2017)
	Nitrogen oxides (SO2),1,4 dichlorobenzene (C6H4Cl2)	1 kg = 1.2 C6H4Cl2 eq	
Terrestrial ecotoxicity	Ethene (C2H4)	1 kg = 4.88×10^4 kg TEG eq. to water	(Owsianiak et al., 2014)

compliance with life-cycle-assessment requirements of Green Star (Green Building Council of Australia, 2014b). Environmental factor evaluation is required by Green Star with seven points for the "Life-cycle impacts" criterion, and 20 points in the "Energy" category.

The model focuses on the variability in building envelopes, required commercial building area, and climate in a different zone in Australia. The proposed model can then be motivated to eliminate difficulties, while green building design can be automated, user-friendly, and effective. Companies can readily use it to implement their green building strategies without the lengthy design processes. The objectives of the model in this part are to calculate the needed energy to balance indoor and outdoor temperature for each climate zone in Australia and to estimate greenhouse gas emissions that are released during the building operations as well as the median energy expenditure in the building lifetime.

The model can help designers choose the type of building envelope by comparison of the reference with the assessed building envelopes (roof, wall, and floor) in Australian regions, which can be chosen in seven climate zones in Australia: Adelaide, Brisbane, Canberra, Perth, Sydney, Hobart, and Melbourne.

The process to input the model parameters is mentioned in Figure 3.7. The outdoor temperature for each location is collected from the Australian Bureau of Meteorology's database. The foundation temperature that connects closely to the model floor is assumed cooler by a mean difference of +3 °C than the air temperature (Gerner & Budd, 2015). The appropriate bounds of office

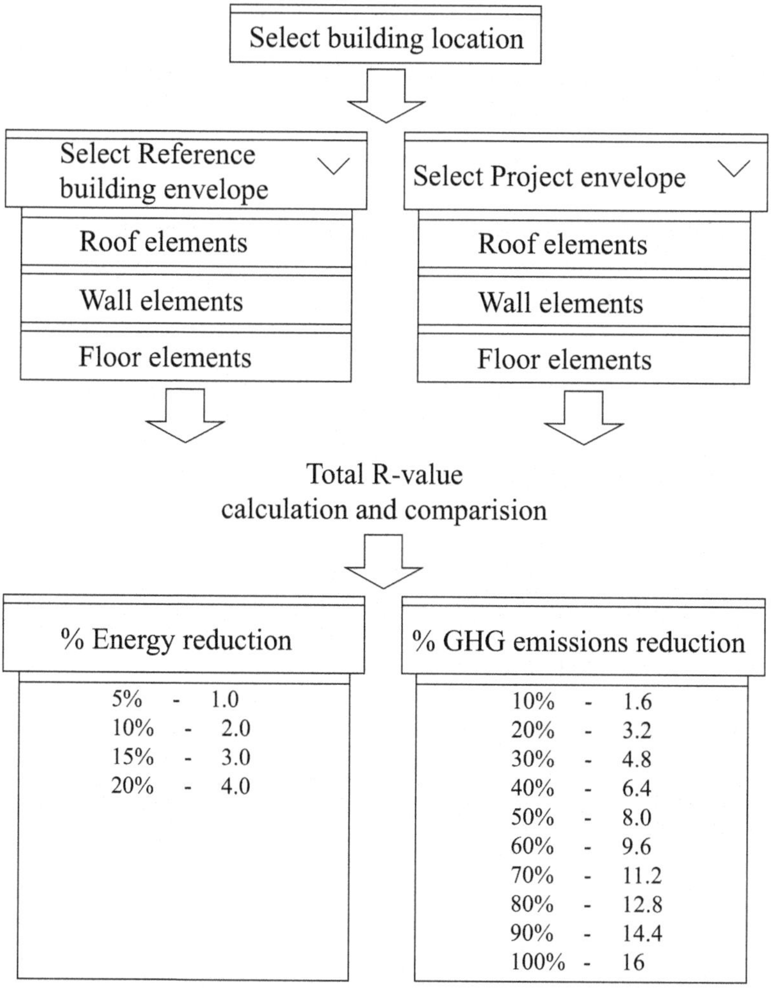

Figure 3.7 Comparison of energy consumption and greenhouse gas emissions

temperature help to improve working quality and reduce negative effects to the employee's attitude. The optimal temperature for office work depends on the season and employee's clothing, and ranges from 20 °C to 26 °C. The internal temperature inside the building used for the model is 25 °C (Victoria WorkSafe, 2008). The number of working hours is 12 h including overtime work. The quantity of energy consumed for heating and cooling the building equals the conductive heat loss rate, which is proportional to the thermal conductance (U-), surface area, and temperature difference, as shown in Equation 3.4 (Jankovic, 2013):

$$Q_c = U \times A \times (T_{in} - T_{ex}) \qquad \text{Equation 3.4}$$

where Qc denotes conductive heat loss rate (W), A is the surface area (m), *T*in is the internal air temperature (K or °C), *T*ex is the external air temperature (K or °C), and *U* is the thermal conductance ($W.m^{-2}.K^{-1}$), which can be calculated by the thermal resistance of a material. The thermal resistance (*R*) is expressed with Equation 3.5 (Butcher, 2015):

$$U = \frac{1}{R}$$ Equation 3.5

The total thermal resistance of the building structure is calculated by Equation 3.6:

$$R = R_{in} + R_{ex} + \sum R_i$$ Equation 3.6

where *R* denotes the total thermal resistance of the structure; *R*in and *R*ex are the thermal resistance of the internal and external surface of the structure; *Ri* is the thermal resistance of physical layers of structure, which is demonstrated in Figure 3.8.

The data of thermal resistance of building elements such as the roof, wall or floor, are collected from the Insulation Council of Australia and New Zealand's handbook (Insulation Council of Australia and New Zealand, 2014) by a model developed in Microsoft Excel. According to this handbook, there are 13 main types of roofing arrangements, 11 types of wall arrangements, and 3 types of floor arrangements.

However, for each of these arrangements, there are several options and combinations for insulation material. For roofing solutions, the required R-value can be obtained by changing different types of sarking materials and ceiling insulation. For walls also, there are wall insulation and membranes. There are mainly ten

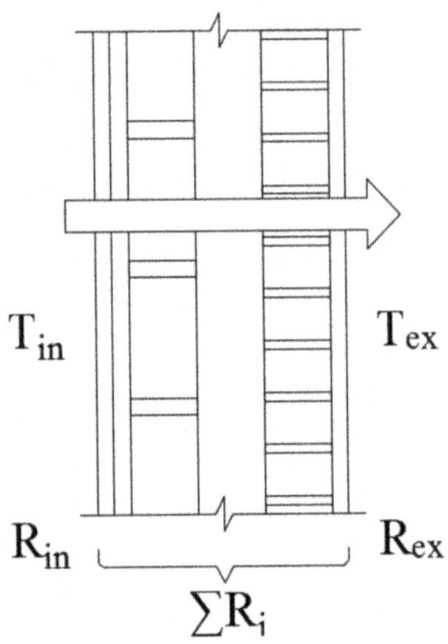

Figure 3.8 Thermal resistance of multi-layer structure

[illegible] products in Australian buildings (Insulation Council of Australia and New Zealand, 2014)

Material *– – Roof elements – -*	*R-value*	*Material* *– – Wall elements – -*	*R-value*	*Material* *– – Floor elements – -*	*R-value*
Tiled roof	0.02	110 mm brickwork	0.18	Timber floorboards	0.12
Concrete slab (150 mm)	0.1	Concrete wall	0.1	Concrete slab (150 mm)	0.1
Metal roof cladding	0	10 mm plasterboard	0.06	Interior air film	0.16
Outdoor air film	0.04	Wall insulation	1.16	Subfloor air film	0.16
Indoor air film	0.11	Wall-vapor permeable	0.19	Ground thermal resistance	0.58
10 mm plasterboard	0.06	Wall-single-side foil	0.2	Floor batts R2.0 (90 mm)	3.1
Rafter remaining air space	0.17	Wall-double-side antiglare	0.71	Floor batts R2.5 (90 mm)	3.6
Nogging air space	0.17	Wall-double-side bubble/foam	0.69	R0.4 EPS board (15 mm)	3.3
Vapor permeable membrane	0	Wall-double-side antiglare EPS	0.88	R0.5 EPS board (20 mm)	3.4
Single-sided foil	0.96	Wall-indoor air film	0.61	R0.7 EPS board (30 mm)	3.7
Double-sided antiglare foil	1.01	Stud wall batts R1.5 (75 mm)	2.5	R1.5 EPS board (60 mm)	4.4
Double-sided bubble/foam foil Rm0.2	0.86	Stud wall batts R2.0 (90 mm)	3	R1.4 board or blanket	2.5
Foil-faced R1.3 blanket	1.05	Stud wall batts R2.5 (90 mm)	3.6	R1.8 board or blanket	2.9
Foil-faced R1.4 blanket	1.21	Stud wall batts R2.7 (90 mm)	3.8	R2.5 board or blanket	3.6
Foil-faced R1.8 blanket	1.59	Wall blanket R1.0 (40 mm)	1.7	R3.0 board or blanket	4.1
Foil-faced R2.3 blanket	2.06	Wall blanket R1.5 (75 mm)	2.2	25 mm air gap	1.8
Foil-faced R2.5 blanket	2.25	Wall blanket R2.0 (90 mm)	2.5	100 mm air gap	1.8
Foil-faced R2.8 blanket	2.72	Wall blanket R2.5 (90 mm)	3	Foil faced board/blanket	0.6
Foil-faced R3.3 blanket	3.01	Wall blanket R2.7 (90 mm)	3.3	R1.4 Reflective PIR (30 mm)	3
R1.3 blanket, 60 mm	1.2	Cement render or plasterboard	0.02	R2.3 Reflective PIR (50 mm)	3.9
R1.4 blanket, 70 mm	1.3	Wall R1.3 blanket	1.7	R3.6 Reflective PIR (80 mm)	5.3
R1.8 blanket, 80 mm	1.7	Wall R1.4 blanket	1.8	R5.5 Reflective PIR (120 mm)	7.1
R2.3 blanket, 100 mm	2.2	Wall R1.8 blanket	2.2		
R2.5 blanket, 100 mm	2.4	Wall R2.3 blanket	2.8		
R3.0 blanket, 130 mm	2.8	Wall R2.5 blanket	3		
R3.3 blanket, 140 mm	3.1	Wall R3.0 blanket	3.5		
R3.6 blanket, 145 mm	3.4	Wall R3.3 blanket	3.8		
Insulation R1.2 (50 mm)	1.2	Wall R3.6 blanket	4.1		
Insulation R1.7 (75 mm)	1.7	16 mm furring channel B2–16	1.7		
Insulation R2.0 (75 mm)	2.1	16 mm furring channel C1–16	2.1		
Insulation R2.0 (115 mm)	2.12	16 mm furring channel C3–16	2.5		

(Continued)

Table 3.10 (Continued)

Material *– – Roof elements – -*	*R-value*	*Material* *– – Wall elements – -*	*R-value*	*Material* *– – Floor elements – -*	*R-value*
Insulation R2.5 (90 mm)	2.6	28 mm furring channel B2–16	1.8		
Insulation R2.5 (140 mm)	2.62	28 mm furring channel C1–16	2.1		
Insulation R3.0 (120 mm)	3.2	28 mm furring channel C3–16	2.4		
Insulation R3.0 (160 mm)	3.22	16 mm channel and foil facing B2–16	2.1		
Insulation R3.5 (185 mm)	3.72	16 mm channel and foil facing C1–16	2.5		
Insulation R4.0 (150 mm)	4.2	16 mm channel and foil facing C3–16	2.6		
Insulation R4.0 (215 mm)	4.22	28 mm channel and foil facing B2–16	2.4		
Insulation R5.0 (180 mm)	5.3	28 mm channel and foil facing C1–16	2.8		
Insulation R5.0 (240 mm)	5.32	28 mm channel and foil facing C3–16	2.9		
Insulation R5.4 (180 mm)	5.7	Air gap 20 mm	2.1		
Insulation R6.0 (260 mm)	6.32	Air gap 35 mm	2.4		
25 mm spacer and air gap	0.43	Sarking material – double-sided bubble /foam foil	0.2		
40 mm spacer and air gap	0.43	Sarking material – double-sided antiglare EPS board	0.38		
55 mm spacer and air gap	0.43				
75 mm spacer and air gap	0.43				
100 mm spacer and air gap	0.43				
Blanket thermal resistance	0.17				

types of insulation used in this model, which include single sided foil, double-sided antiglare foil, vapor permeable membrane, bubble/foam foils, ceiling and wall batts, foil-based blanket, foil faced board, antiglare expanded polystyrene (EPS) board, and reflective rigid polyisocyanurate (PIR) sheets (Insulation Council of Australia and New Zealand, 2014). The R-values of these insulation products are shown in Table 3.10.

The environmental impact result of 1 MJ of energy is extracted from the model developed in GaBi 8.7 (Figure 3.9). The model is designed to be flexible and is one of the simplest ways in choosing the optimal building envelope for achieving Green Star energy credit when focusing on insulation. This model can identify the most suitable type of insulation for the selected structure in the selected climate zone. Apart from that, this provides the greenhouse gas emission calculation for the selected option. Thus, this can be used to compare the reference building and the assessed project with both the results of energy consumption and greenhouse gas emissions. This model could become an independent application and consequently can be applied to different building configurations.

3.4.2.3 *Life-cycle greenhouse gas emissions assessment for concrete in relation to Credit 19*

Cement is the fundamental element in concrete. Its production is responsible for approximately 5% of the global industrial energy consumption and 5% of the global CO2 production (Horvath, 2004). In this simulated model, concrete has mainly been categorized into two types: conventional and high-strength concrete. Concrete made with normal aggregate with compressive strengths greater than 40 MPa (~6000 psi) is considered high-strength concrete, while concrete

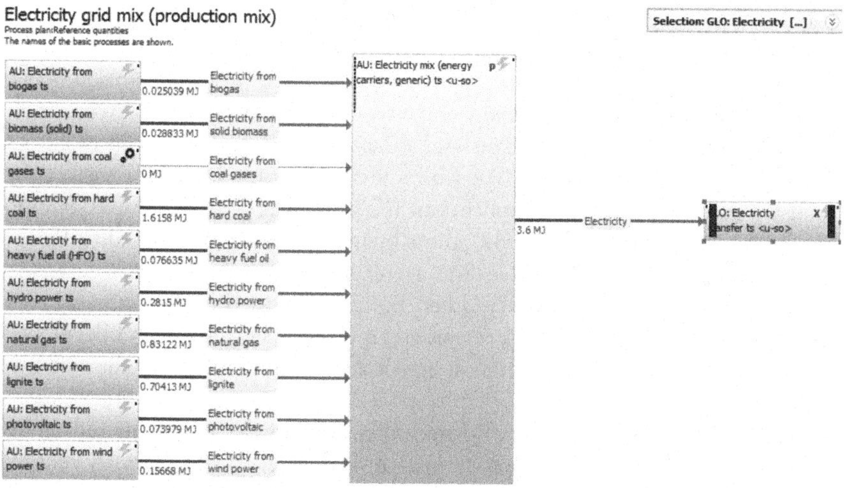

Figure 3.9 Model for electricity grid mix – (a screenshot of the model flow in GaBi 8.7)

with compressive strengths lower than 40 MPa is considered as conventional concrete (Mamlouk & Zaniewski, 2011).

Life-cycle greenhouse gas emissions from concrete production are directly proportional to the cement content used in the concrete mix (Sabnis, 2015; Tam et al., 2018b). It is also well-known that among the principal materials used for concrete manufacturing, greenhouse gas emissions are typically attributed to the Portland clinker production from cement kilns. Most cement contains approximately 5% of gypsum, 12% of supplementary cementitious material including fly ash (or pulverized fuel ash), superfine fly ash, ground granulated blast furnace slag, rice husk ash, natural pozzolans, colloidal silica, metakaolin, superfine Calcium Carbonate (pure limestone), and 83% of Portland clinker (Kumar, 2010; Day et al., 2013).

A variety of sustainable software exists that professionals and researchers use nowadays, such as Gabi, SimaPro (Speck et al., 2016). However, the existing tools do not specifically cater to life-cycle greenhouse gas emissions assessment for Green Star credits. Thus, it is a time-consuming process to effectively use these programs for achieving Green Star status. Meanwhile, Microsoft Excel is considered to be the most common and familiar software to designers and quantity surveyors (Eastman et al., 2011). To help project parties be able to achieve Green Star status in particular, the primary objective of this book is to develop an MS Excel-based model to examine life-cycle greenhouse gas emissions for concrete implementation and explore adequate calculations and results to successfully achieve full credit points for Credit 19: Life-cycle impacts – Concrete for the Green Star – Design & As Built under the Green Star Environmental Rating System.

This chapter presents a systematic approach to Credit 19 and addresses all concrete types used in a project, including structural and non-structural elements as well as the appropriate proportions of supplementary cement materials used in concrete (Green Building Council of Australia, 2015a). Different supplementary cement material types are commonly incorporated into the concrete mixture as substitutes for fly ash, silica fume, or blast furnace slag and this method is particularly effective from the economic and environmental perspective (Abbas et al., 2006). The researched mix designs include various concrete strengths of 20, 25, 32, 40, 50, 65, 80, and 100 MPa, which are mainly used in the construction industry for several different applications (Kosmatka et al., 2002; ACI Committee, 2008). Achieving extra credit points by incorporating recaptured/reclaimed water and the use of alternative fine aggregate within the mix has also been explored (Kosmatka et al., 2002; Green Building Council of Australia, 2015a). The benefits of conducting this research can provide businesses and developers access to various concrete mixture designs that can achieve full credit points in Credit 19.

Concrete mixture design is also necessary to provide an automated model rather than manual calculations because the data given are frequently referred to when intermediate values are required to achieve the available credit points in Green Star concrete credit. Concrete mixture design, which is formed from a process consisting of two interrelated steps: (1) selection of the proper ingredients (including cement, aggregate, water and admixtures) of concrete; and (2) determining their relative quantities ("proportioning") to be as most economical as possible.

Concrete mixture design for different concrete strengths can be estimated using the steps shown in Figure 3.10 (Kosmatka et al., 2002; Mindess et al., 2003):

(1) Selection of concrete strength grade: it is necessary to collect the required information on required materials and structure elements.

(2) Selection of the appropriate water-to-cement ratio or water-to-SCM ratio for ordinary concrete, or water-to-binder ratio for the high-strength concrete, which is not governed by concrete strength but by durability requirements (based on exposure conditions of concrete): Concrete strength is mainly determined by the water-cement ratio (Hassoun & Al-Manaseer, 2012). A water-cement ratio is about 0.45 or higher for concrete with compressive strengths lower than 35 MPa. A water-cement ratio of 0.45–0.35 is necessary for the concrete with 35–55 MPa compressive strength, respectively (Caldarone, 2009). To produce high-strength concrete with specific cement content requires both a low water-to-SCM ratio and low water content. Water content for high-strength concrete should be lowered to under 145 kg/m^3, and preferably be in the 125–135 kg/m^3 range (Mindess et al., 2003).

(3.1) Selection of appropriately required slump: for both conventional and high-strength concrete types, the assumed slump range in this model is 30–50 mm.

(3.2) Selection of coarse aggregate type and size: the compressive strength of high-strength concrete generally decreases as coarse aggregate size increases. Concrete with compressive strengths up to 60 MPa may use coarse aggregate with a maximum size of 19 or 25 mm. The proposed computer-aided model can estimate the aggregate with a maximum size of 9.5 or 12.5 mm for both regular and high-strength concrete (Mindess et al., 2003).

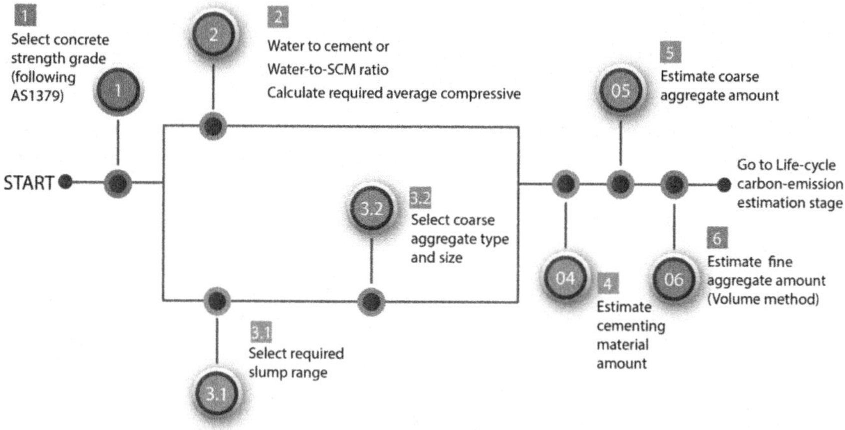

Figure 3.10 Flowchart representation of the concrete mixture design estimation

(4) Calculation of cement or SCM content: using water content and water-to-cement or water-to-binder ratios, cement content is inversely proportional to the water-to-binder ratio (Caldarone, 2009). Proper mixtures with typical cement content for high-strength concrete are ranging from 390 to 560 kg/m^3 including the use of superplasticizers, and high coarse-aggregate content can result in the production of high-strength concrete (Mindess et al., 2003).

(5) Estimation of the coarse-aggregate content: by obtaining higher strength levels with smaller maximum aggregate sizes in the 9.5–12.5 mm range (Kosmatka et al., 2002) usually yields better concrete performance and economy. The proposed model assumes that the maximum aggregate size for both conventional and high-strength concrete is 12.5 mm.

(6) Estimation of the content of fine aggregate: Using a proposed volume method in the computer-aided model, the required content for fine aggregate can be estimated.

(7) Adjustment for moisture in the aggregate.

Greenhouse gas emissions are primarily related to material and energy flow within the boundary of the building's life-cycle, mainly during the following processes: extraction, manufacturing, on-site operation, demolition, recycling, disposal, and transportation (Nielsen, 2008; Linfei et al., 2011; Gong & Song, 2015; Zhang & Wang, 2015). Greenhouse gas emission quantity includes emissions from the extraction/production to the disposal stage using an emission-factor approach that is calculated by each emission factor's rate along with these activities' parameters (Interstate Natural Gas Association of America, 2005; Chau et al., 2015; Seo et al., 2015).

A GREENHOUSE GAS FOOTPRINT OF CONCRETE EXTRACTION AND PRODUCTION PHASES

Dominant greenhouse gas emission sources generally lie in the process of building material production, construction waste processing, consumption of fossil fuel-generated energy, machinery, equipment production, and transportation. The six categories are identified as embodied carbon in material, construction processes, energy production and management, transportation, and water and waste management (Chunxia et al., 2010; Bunning et al., 2013; Thomson et al., 2013).

Greenhouse gas emissions generated from building material production and construction waste processing are estimated by summation over a set of multiplication of greenhouse gas emissions produced by material (i) (in kg) and global warming potential (GWP) of greenhouse gas emissions over a 100-year period (Fan & Long, 2009).

The model of this book does not include any greenhouse gases other than CO2. The GWP for carbon dioxide over a 100-year interval amounts to 1 (Houghton, 1996). Greenhouse gas emissions for the concrete extraction and production phases can be expressed in Equation 3.7:

$$GHG_{\text{Extraction/Production}} = \sum_{i}^{n} EC_i \times Q_i \qquad \text{Equation 3.7}$$

where GHGExtraction/Production represents the amount of greenhouse gas emissions during the material extraction process and concrete production, EC*i* denotes EC emission data of the *i*th type of material in concrete mixture (in kg CO2-eq/kg) (Hammond & Jones, 2008), and *Qi* is the quantity of material (*i*) used in the concrete mixture design, which is estimated in Section 2.2 (in kg).

B GREENHOUSE GAS EMISSIONS FOR TRANSPORTATION AND PUMPING PHASES

The calculation of greenhouse gas emissions from the concrete transportation process to the construction site includes the emissions released from a concrete pump machine and mobile concrete mixing truck. The amount of energy usage for raw material and their final products transportation can be estimated by multiplying the average distance within a construction site and the average energy consumption of involved vehicles (Nielsen, 2008; Morrissey & Horne, 2011). Equation 3.8 shows the estimation of greenhouse gas emissions generated in this phase (Zhang & Wang, 2015):

$$\mathrm{GHG}_{\mathrm{transportation}} = \sum_{i}^{n} Q_i \times \mathrm{AD} \times \mathrm{GHG}/\mathrm{km} \qquad \text{Equation 3.8}$$

where GHGtransportation denotes the amount of greenhouse gas emissions of the transportation for raw material (in kg), AD is the average transport distance (in km), and GHG/km denotes the resultant greenhouse gas emissions released by the vehicles (in kg CO2-eq/kg material/km). Embodied emissions for loading vehicles widely vary from 0.059 to 0.109 × 10^{-3} kg CO2-eq/kg and can be averaged to 0.1 × 10^{-3} kg CO2-eq/kg (McKinnon & Piecyk, 2010). The assumed transportation distances should be re-assessed for each case of project and concrete supply approach. Maintaining the best concrete workability condition involves working periods of around 1 to 2 hours in case concrete arrives on site at the 80–90 mm slump (Department of Transport and Main Roads, 2014). Speed limits for heavy vehicles in Australia range regularly from 10 km/h in shared zones to 100 km/h on open roads (Katie Willis and Simone Gangell, 2012).

Thus, the average distance travelled is assumed to be 50 km (Zhang & Wang, 2015). The amount of greenhouse gas emissions from the pumping stage is estimated by multiplying an emission factor by fuel consumption amount of diesel engine types specified in the National Pollutant Inventory Emission Estimation Technique Manual for Combustion (Lawrence et al., 2005; Australian Government, 2008). A 100-kW pump engine was adopted with a flow rate of 20 m^3 concrete, corresponding to the energy consumption of 5 kWh/m^3 of concrete (Crossin, 2012). The emission factor for diesel is 0.305 kg CO2-eq/KWh (Covenant of Mayors, 2016). The fuel consumption for the 100-kW pump engine (with the consumption rate is 0.18 kg/(hp × h) and one mechanical horsepower = 0.745 kW, working for 2 hours, is adopted as the best concrete workability period, corresponding to 0.241 kg/kWh [= 0.18 kg/(hp × h) /0.745 kW)] (Department of Transport and Main Roads, 2014; Ma et al., 2016).

C GREENHOUSE GAS EMISSIONS FOR THE DEMOLITION PHASE

The construction and eventual demolition of concrete structures requires a large number of resources and energy, and as a result, places a burden on the environment (Japan Society of Civil Engineers, 2006). Equation 3.9 is used to calculate the greenhouse gas emissions generated during this phase. The greenhouse gas emissions of 0.008 kg CO2-eq /kg (Nielsen, 2008) is commonly used as a reference for the demolition phase because the proposed mixes used in the calculations are not heavily reinforced (Zhang & Wang, 2015):

$$\mathrm{GHG}_{\mathrm{demolition}} = D \times Q \qquad \text{Equation 3.9}$$

where GHGdemolition denotes the amount of greenhouse gas emissions released during the demolition phase, *D* denotes the demolition emissions released during the process (kg CO2-eq /kg), and *Q* is the concrete quantity to be demolished (in kg).

D GREENHOUSE GAS EMISSIONS IN THE DISPOSAL PHASE

The landfill method is a major method of waste treatment in Australia, including domestic as well as industrial waste (Australian Bureau of Statistics, 2013). Greenhouse gas emissions at the disposal phase can be calculated in a similar manner to the transportation phase. Equation 3.10 is used to calculate the amount of greenhouse gas emissions in the disposal phase (Nielsen, 2008):

$$\mathrm{GHG}_{\mathrm{disposal}} = Q(\mathrm{AD})(\mathrm{GHG}/\mathrm{km}) \qquad \text{Equation 3.10}$$

where GHGdisposal denotes the amount of greenhouse gas emissions caused during the disposal phase for concrete waste (kg CO2-eq/kg), and Q is the waste material (in kg). A truck for disposal transportation emits about 100×10^{-6} kg CO2-eq/km (Nielsen, 2008). The average distance (AD) travelled can vary among projects, and this model assumes that the average distance under the proposed model is 50 km.

E GREENHOUSE GAS EMISSIONS FOR THE RECYCLING AND REUSE PHASES

The concrete industry practically incorporates various environmental management practices that comprise waste reuse and recycling from concrete products including water and unused concrete (Lemay, 2011). The reuse and recycling industry can significantly lower the greenhouse gas emissions impact of buildings on the environment (Australia Government, 2012). The most popular method to recycle concrete is crushing demolished concrete waste to generate recycled aggregate for concrete production (Tam & Tam, 2008). Concrete reprocessing involves the use of relatively uncomplicated and well-established crushing techniques (Australia Government, 2012).

A factor of concrete recycling emission is made up of two components: process energy and transportation energy (Environmental Protection Agency, 2015a). In

Table 3.11 State-wise recycling rates in Australia for 2010–2011

State	*Industrial recovery rate*
Tasmania	27%
Victoria	66%
New South Wales	63%
Queensland	46%
South Australia	81%
Average rate	**58%**

a report from the Commonwealth Scientific and Industrial Research Organisation (CSIRO) for the building industry in 1999, a recommendation noted that a 30% recycled aggregate replacement could be used for construction (Australia Government, 2012). The concrete recycling rate depends on many factors including, but not limited to, the recovery cost and technologies (Tam & Tam, 2008).

Table 3.11 shows that the industry has achieved the average recovery industrial recycling rate of about 58%, and this rate has been steadily rising in Australia (Blue Environment Pty ltd, 2014). The recycling process with well-established and straightforward techniques is mobilized for construction waste such as masonry materials, concrete, and bricks. Concrete waste to be recycled occupies about 60% of reprocessing masonry materials (Christophe Brulliard et al., 2012). An assumption of 35% disposal quantity (60% × 58%) is used for Equation 3.11:

$$GHG_{recycling} = 35\% GHG_{disposal} \qquad \text{Equation 3.11}$$

where GHGrecycling denotes the greenhouse gas emissions amount released during the recycling phase of each material, and GHGdisposal denotes the amount of greenhouse gas emissions released during the disposal phase (in kg CO2-eq /kg). Emission factors data for concrete material and production stage can be referenced in Table 3.12.

The concrete mixture is developed in the model with some assumed characteristics: (1) strengths within the 20–100 MPa range; (2) aggregate types as angular coarse (crushed stone), sub-angular aggregates, gravel with some crushed particles and rounded gravel; and (3) exposure to environmental condition of the concrete (Bickley et al., 2006; American Concrete Institute, 2014). Figure 3.11 illustrates the developed model in GaBi 8.7 for 1 m^3 20 MPa concrete with 30% fly ash, 50% water reduction, and 25% manufactured sand, which include the mixture element quantity conducted from the model calculation in Excel and Visual Basic.

3.4.2.4 *Steel product and Green Star Credit 19B*

Steel is the most extensively used metal construction material globally (Andersson, 2013). Therefore, it is especially crucial to conduct a quantitative environmental impact study on this material. There are two points available to be achieved in

Table 3.12 Emissions factors for concrete materials

No.	*Concrete materials*	*Embodied carbon EC kg CO_2-eq/kg*	*Description*
1	Water	0	
2	Ordinary Portland Cement (OPC)	0.82	The most common cement used in concrete construction, which is best suited where there is no exposure to groundwater or sulfates in the soil (Neville & Brooks, 2010; Australian (Iron & Steel) Slag Association, 2012).
3	Coarse aggregate	0.0046	A collection of materials including stone, gravel etc. that generally ranges from 9.5 to 37.5 mm in diameter (Collins, 2010; Australian (Iron & Steel) Slag Association, 2012; American Society for Testing and Materials, 2013).
4	Fine aggregate	0.0014	Natural sand or crushed stone with mostly particles (Australian (Iron & Steel) Slag Association, 2012; American Society for Testing and Materials, 2013).
Supplementary cement material			
5	Fly ash	0.027	A fine gray powder made up of mostly spherical glass-type products which is a by-product produced by coal-fired power stations (Flyash Australia, 2010).
6	Blast furnace slag	0.143	A by-product of the manufacture of iron, which has the properties of a cementitious binder (Australian (Iron & Steel) Slag Association, 2012; Green Building Council of Australia, 2015a).
7	Silica fume	0.014	A powdered by-product created in the production of silicon metal or ferrosilicon alloys (Cement Concrete & Aggregates Australia, 2010; The Silica Fume Association, 2016).
8	Alternative fine aggregate	0.00095	Manufactured sand: A purposely made crushed fine aggregate produced from a suitable source material and designed for use in concrete or other specific products (Cement Concrete & Aggregates Australia, 2007).

Credit 19B, where a reduction of steel mass is demonstrated in comparison with standard practice. The process to estimate the environmental impact of steel is similar to the process to the impact estimation of other materials mentioned in Sections 3.4.2.1 and 3.4.2.3. The life-cycle greenhouse gas emissions assessment of this material is also implemented for the production, transportation, construction, demolition, recycling, and disposal phases. The life-cycle greenhouse gas emissions assessment is conducted using GaBi 8.7 with the functional unit for the analysis of 1 kg steel in 50 years of the building's life. Equation 3.3, which calculates the impact category indicator, is also used to calculate the life-cycle impact of steel used in the building.

Energy sources for producing primary steel use about 50% coking coal, 35% electricity, 5% natural gas, and 5% other gasses (Guggemos Angela & Horvath, 2005). The recycling rate of construction steel is approximately 85% of the quantity collected from destruction processes (Bird, 2018). In this section, the assessment of life-cycle greenhouse gas emissions for steel illustrated in Figure 3.13 also includes the same impact categories with the materials described in previous sections, which are referred to in Table 3.8.

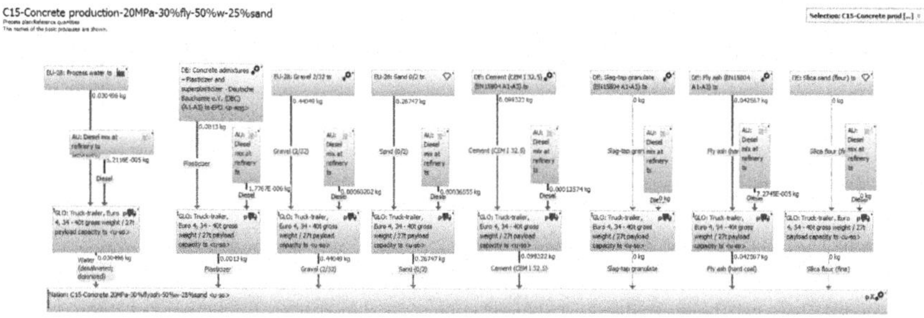

Figure 3.11 Model for 1 kg 20 MPa concrete with 30% fly ash, 50% water reduction, and 25% manufactured sand – (a screenshot of the model flow in GaBi 8.7)

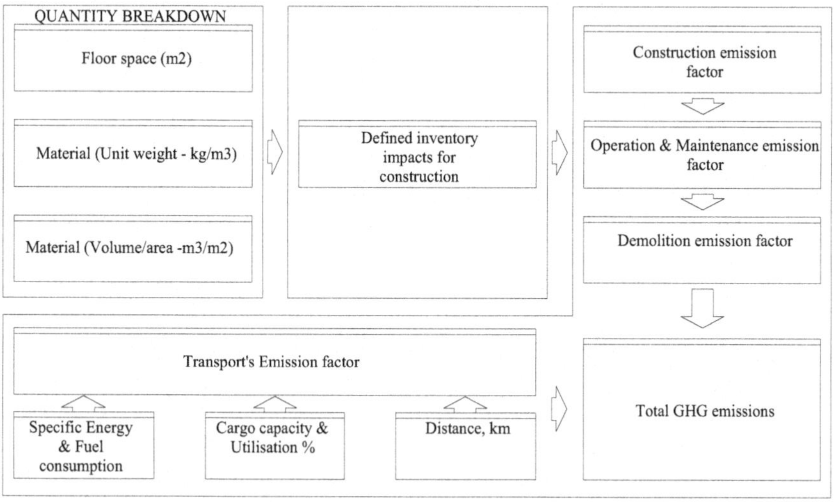

Figure 3.12 Greenhouse gas emissions calculation flowchart for steel life-cycle assessment

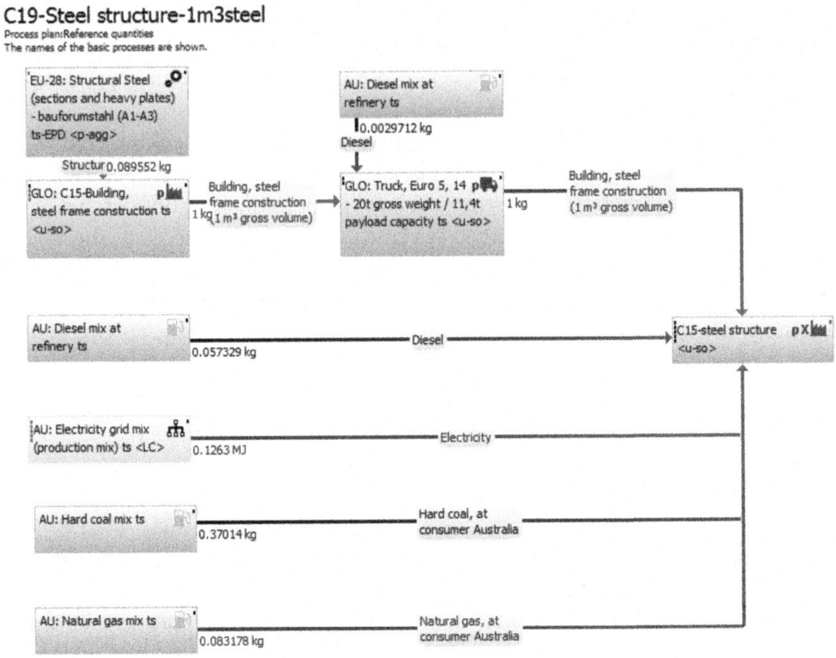

Figure 3.13 Model for 1 kg steel – (a screenshot of the model flow in GaBi 8.7)

3.5 Model interface

The input process of the model is gradually implemented following the steps portrayed in Figure 3.14, Figure 3.15, and Figure 3.16. As shown in Figure 3.14, the first step is the selection of the building's location in order to identify the outdoor temperature of the project. The envelope module of the model is expressed in Figure 3.15. In this module, the building fabric options of the reference building and assessed project are selected to equate the R-value of each layer of the structures using MS Excel. The program then calculates the total R-value mentioned in Equation 3.6. The quantity of energy consumed for calibrating the temperature in the building is calculated by Equation 3.4. The environmental impacts of one unit of energy are extracted from the model run in GaBi 8.7 (Figure 3.9).

The materials module of the model is expressed in Figure 3.16. Timber and reinforcement bar quantity are input in the table as shown in Figure 3.14. The environmental impacts of these materials are assessed with the process in Figure 3.6 and Figure 3.13. Because there is a variety of combinations of concrete mixture with Portland cement, supplementary cementitious materials, reclaiming 50% the amount of water, and alternative coarse or fine aggregates, the Visual Basic code to automatically select the type of concrete and export the results is presented as the procedure shown in Figure 3.16.

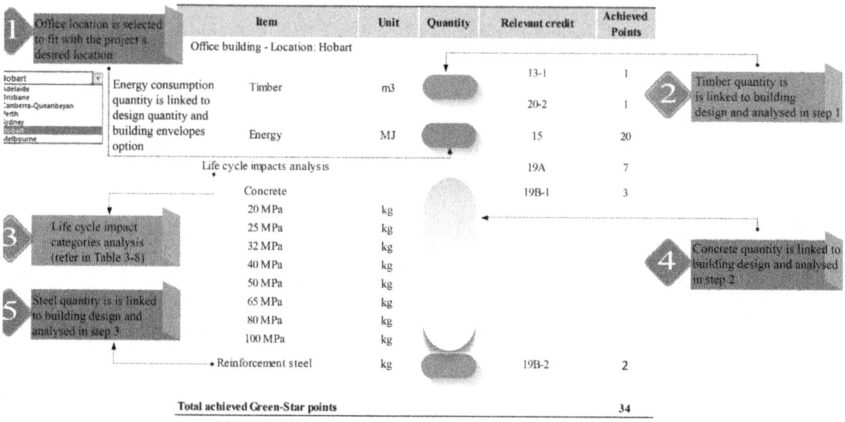

Figure 3.14 Main input interface of the model

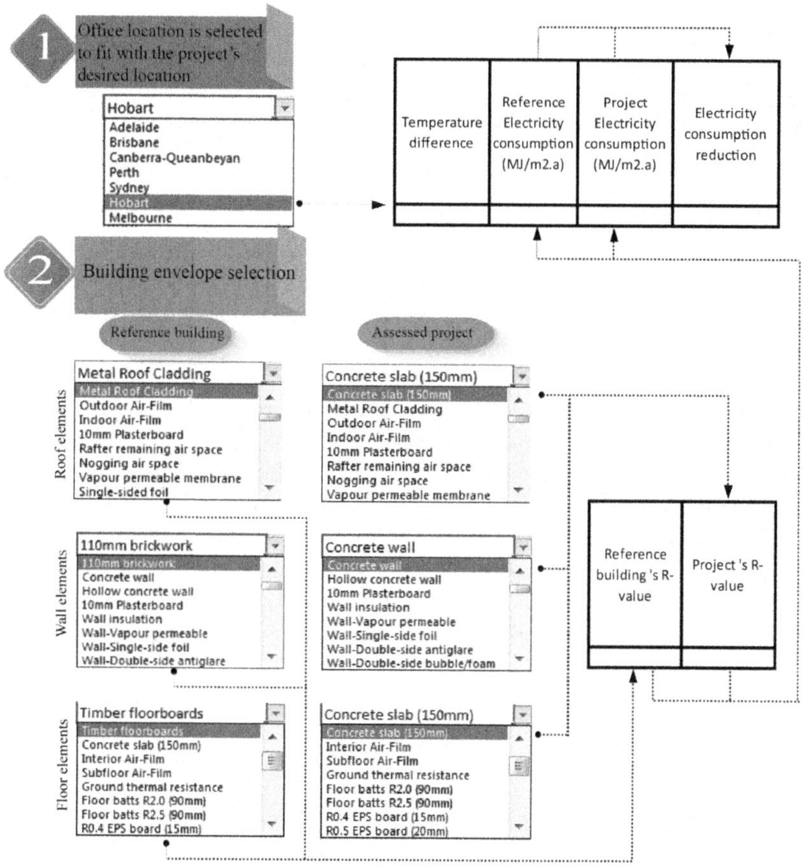

Figure 3.15 Envelope module of the model

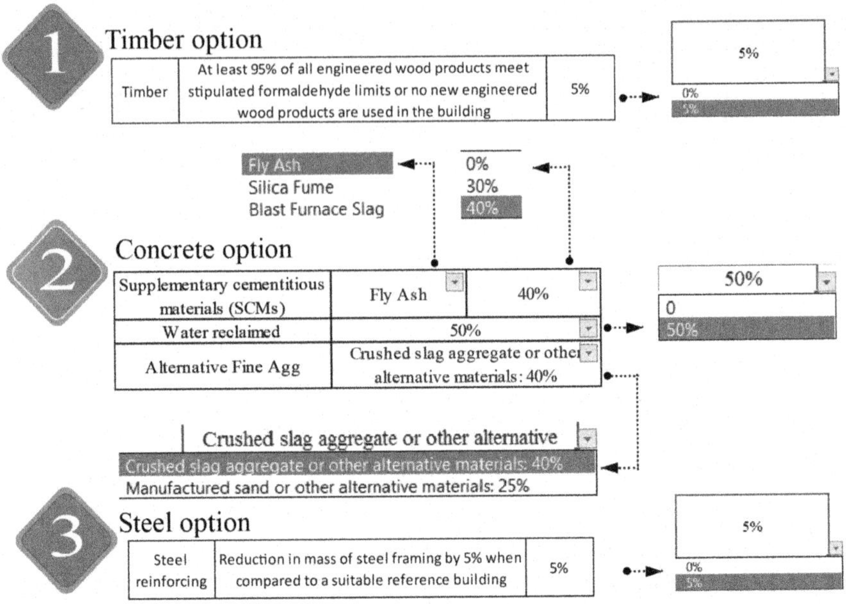

Figure 3.16 Materials module of the model

3.6 Summary

This chapter presents the process to estimate and assess the life-cycle greenhouse gas emissions for the building elements. The credits in Green Star – Design & As-Built, which are related to environmental impact during the building's life-cycle, are discussed in this chapter.

The chapter provides terminological discussions on the credits link with greenhouse gas sources in the Green Star rating tool, and the credit points that can be achieved by the developed model. Furthermore, the chapter discusses the role of the Green Star rating tool in the process of life-cycle assessment and the categories, as well as credit points allocation of this tool. After the review of Green Star credits and the intersection between the tool and the life-cycle greenhouse gas emissions reduction, this chapter illustrates the chosen credits and the building elements to be analyzed and applied in the research model, which will estimate life-cycle greenhouse gas emissions and select the optimal design to achieve the goal of a green building.

4 Environmental impacts on the model's parameters

4.1 Introduction

This chapter comprises two major discussion parts that investigate the environmental impact results of all parameters linked with the greenhouse gas sources reduction credits in the Green Star rating tool mentioned in Chapter 3. The first part provides the discussion on the environmental impact of building envelopes in commercial buildings in a variety of geological zones in Australia, in order to reduce energy consumption and greenhouse gas emissions. The second section examines the key materials often utilized in such a construction. These chapter parts are discussed in terms of the credit points options in Green Star that a green building can achieve. Then, the models' parameters are discussed with other green building assessment tools, for example, LEED, BREEAM, etc., in order to compare how designers can achieve green certificate for their projects with other tools rather than the Green Star system.

4.2 Environmental impacts of commercial building envelopes

The model results show potential discrepancies among life-cycle-assessment results from different methods. These proposed analyses have provided guidance to life-cycle-assessment practitioners, who can select the project design depending on specific approach and impact categories. The global warming potential assessment is conducted to measure the average energy consumption per 1 ton of carbon dioxide gas released into the atmosphere (Owsianiak et al., 2014). Among these methodologies, impact results that were estimated using ReCiPe 2016 are the highest and those obtained using I02+ are the lowest. The analyses of the model for global warming potential impact show that the structures utilizing concrete contribute at a higher level to the global warming potential, which was estimated using ReCiPe, ILCD, TRACI, and I02+ methodologies. The global warming potential results mostly from structures using brick and concrete. The impact from structures that are made from timber contributes a small portion to the total global warming potential impact, on average (Bueno et al., 2016). Results of this impact category are slightly different among these life-cycle-assessment methodologies and have a similar trend to the total impact results.

The ozone layer depletion is defined as the ratio of the ozone breakdown due to given substances to the ozone breakdown due to an equivalent quantity of CFC-11 (Heijungs, 1992). Ozone depletion potential results are different among the utilized methods. Substance impact results are quite similar for ILCD, TRACI, and I02+ categories; however, analyses using the ReCiPe method yield significantly higher calculated impact than other methods. The results obtained using ILCD, TRACI, and I02+ methods are considerably lower than those using the ReCiPe method.

The stratospheric ozone depletion results using the ReCiPe method are the most effective among the four approaches: ReCiPe, ILCD, IMPACT 02, and TRACI. Nitrous oxide (N2O) is the primary substance that contributes to ozone depletion potential impact in the ReCiPe analyses. The considerable difference between the ReCiPe method and others is due to a divergence in substance coverage, as carbon tetrachloride (CCl4) is mainly recognized among an emission group hierarchy in ILCD, IMPACT 02, and TRACI, while N2O is the significant contribution in the ReCiPe method. CCl4 impact scores are quite small and do not change much when the energy consumption reduction scenarios are used. However, with the ReCiPe 2016 approach, it appears that N2O results will inevitably decrease when the model reduces the energy consumption.

As opposed to the TRACI method, which employs maximum incremental reactivity covering most substances and explicitly used in the USA, the use of ReCiPe, ILCD, and I02+ for photochemical ozone impact is more widespread in other countries (Bare, 2011). However, the photochemical ozone creation potential results express the inconsistency among these three methodologies. This is probably because ReCiPe 2016 is replacing a European factor employed in ReCiPe 2008 by an average international factor, which is based on regional-specification factors (Huijbregts et al., 2017). Meanwhile, the ILCD method uses the LOTOS-EUROS model to interpret photochemical ozone creation potential impact on human health, and the I02+ method assesses the impact factors by classifying the damage factor collected directly from the Eco-indicator 99 (Goodkoop, 1999; Van Zelm et al., 2008).

For aquatic ecotoxicity, 1,4-dichlorobenzene equivalents (1,4-DB eq) to ReCiPe, or comparative toxic unit (CTUe) to ILCD and TRACI, or triethylene glycol equivalents (TEG eq) to the I02+ were chosen as the reference substance (Owsianiak et al., 2014). While result patterns by ReCiPe show that the structure that uses metal as the building envelope has a higher level of aquatic ecotoxicity (1.19×10^{-1}, 1,4-DB eq), the application of ILCD, TRACI, and I02+ methodologies suggests that the structure using steel contributes to the leading concentration. This is because the impact result of analysis of a metal is higher than concrete in the ReCiPe method than in the other methodologies. Aquatic ecotoxicity results using these selected methods also appear difficult to compare because different characterization model approaches were employed (Owsianiak et al., 2014).

A complete inventory of substances connected to the acidification potential impact factor, including hydrogen fluoride (HF), hydrogen sulfide (H2S), and sulfur dioxide (SO_2), is expressed in terms of their standard substance SO_2 (Heijungs, 1992). The considerable impact percentage of brick masonry accounts for approximately 60% of the assessed emissions. The substantial contribution is from a concrete slab in the roof and floor structures, which accounts for about 20% of the contributing substances.

The human-toxicological effect analyses using ReCiPe method are stated in terms of benzyl and methylbenzene equivalent emissions to air, and used the reference unit, kg 1,4-DB eq (Hertwich et al., 2001). Impact results of masonry structures are higher than the other options. An impact indicator of the human toxicity potential calculated by ILCD and TRACI methods is the comparative toxic unit for human (CTUh), which expresses the estimated increase in morbidity among the total human population per unit mass of a substance released (cases per kilogram). In these methods, half of the emission contributions are significantly realized from electricity sources. This result clearly demonstrates the necessity to reduce energy consumption during the operational phase, aiming to reduce the risk to human beings. The results show this book's ability to adopt various options for building envelopes, which are included under the National Construction Code of Australia and the handbook of Insulation Council of Australia and New Zealand (2014), and energy consumption scenarios in Green Star. This model compares results from the global warming potential or climate change category life-cycle greenhouse gas emissions results, estimated using ILCD and TRACI methods, which appear quite similar. Nevertheless, results obtained using the ReCiPe method appear the highest. The estimated environmental impact using the IMPACT 2002+ methodology appears the lowest among these four methodologies.

Most of the released substances stem from material production stages and their usage processes during the construction phase. These results indicate that the use of brick and concrete in buildings severely impacts the environment. Life-cycle greenhouse gas emissions would increase when structure material is changed from the combination of tiled roof, one-layer brick wall, to the envelope with metal roof and two-layers brick wall.

The contributed environmental impact of electricity consumption to life-cycle greenhouse gas emissions in the building life-cycle depends on the decision of choosing brick or concrete for the building envelope (Kua & Kamath, 2014; Tam et al., 2017d). The smaller the total R-value that the model has, the higher the energy that the building would consume, leading to higher greenhouse gas emissions during the building's life-cycle. Life-cycle greenhouse gas emissions estimated from the combination of concrete roof and floor with masonry wall structure are higher than life-cycle greenhouse gas emissions calculated by their combination, which employs only concrete structures. By changing the energy consumption scenarios in the building that comply with Green Star Credit 15 requirement, the life-cycle greenhouse gas emissions changing trend also has a positive relationship with the amount of energy consumed.

For acidification analyses, NH_3, NO_x, and SO_x are common substances, which are recognized as triggering acidification impacts in land ecosystems. Because

IMPACT 2002+ combines terrestrial eutrophication and terrestrial acidification impacts, the resultant impact scores appear much higher than results obtained using other methodologies (Seppälä et al., 2006; Owsianiak et al., 2014; Huijbregts et al., 2017). With the proposed analyses, energy consumption reduction can be realized as an important criterion for reducing the model's life-cycle greenhouse gas emissions using all four methodologies.

Using the proposed human toxicity approach, the quantity of heavy metal and organic emissions to the atmosphere severely impacts the environment. The cancerous effects, according to ReCiPe, are caused by arsenic (+V), lead, and formaldehyde emitted to the atmosphere, while, according to TRACI and ILCD, mercury and chromium are the cause of the impact. There are indicators of human toxicity potential, which are expressed using the reference unit of kg 1.4-DB eq in ReCiPe 2016, and CTUh in LCD 2009 and TRACI 2.1. The environmental impact of electricity consumption reduction, estimated using these methods, is also apparently recognized.

These results show the ability of the study to adapt to the Green Star "Energy" scenarios for energy consumption reduction and the environmental effects during the building's life-cycle. Eight environmental impact categories mentioned in Section 3.4.2 meeting the requirements in the "life-cycle impacts" criteria are also assessed using the model for each scenario. Using this model, life-cycle assessment practitioners can compare and select the optimal sustainable impact design, and hence achieve Green Star certificates for their projects. Table 4.1 shows that the amount of energy consumption needs for a reference building and an assessed building example is calculated following the choice of elements used for the buildings. Table 4.1 shows the comparison of the project envelope options with total R-values collected from Table 3.10. The reduced percentage of energy consumption and greenhouse gas emissions is about 95%. Compared with the criteria in Credit 15, the points achieved for the assessed example are 4 points for energy consumption reduction and 16 points for greenhouse gas emission reduction.

4.3 The impacts of materials for green buildings

4.3.1 Application of the reduction of timber and steel in green buildings

The results of the life-cycle impacts are used the inventory data, which are based on the Australian Life Cycle Inventory library and the life-cycle assessment approach of CML2001, ILCD, ReCiPe 2016 and TRACI 2.1 in GaBi 8.7 (Figure 3.6 and Table 4.2). The reduced environmental impacts are equal to 5% of the original environmental impacts, demonstrating that the project deploys 95% of recycled and/or reused timber products. By applying these criteria of Credits 13 and 20, the project successfully obtains 2 points to add to the total.

The reduced environmental impact results are equal to 5% of the original environmental impact results, demonstrating that the project reduces 5% of steel framing when compared to an appropriate reference building (Table 4.3).

Table 4.1 Comparison of life-cycle greenhouse gas emissions between reference building and assessed project

Impact category	*Life-cycle assessment method*	*Unit*	*Unit environmental impact for 1MJ*	*Reference building environmental impact*	*Green Star environmental impact*	*Reduced environmental impact*	*Reduced percentage*
Global warming potential (GWP 100 years)	CML2001 – Jan. 2016, global warming potential (GWP 100 yrs.)	kg CO_2 eq.	0.2453	351.10	16.55	334.55	95.29%
	I02+ v2.1 – Global warming 500 yrs. – Midpoint	kg CO_2 eq.	0.2385	341.48	16.09	325.38	95.29%
	ILCD – Climate change midpoint, incl. biogenic carbon (v1.09)	kg CO_2 eq.	0.2445	350.01	16.49	333.51	95.29%
	ReCiPe 2016 Midpoint (H) – Climate change, incl biogenic carbon	kg CO_2 eq.	0.2376	340.16	16.03	324.13	95.29%
	TRACI 2.1, Global Warming Air, incl. biogenic carbon	kg CO_2 eq.	0.2445	349.92	16.49	333.43	95.29%
Aquatic ecotoxicity	CML2001 – Jan. 2016, Freshwater Aquatic Ecotoxicity Pot. (FAETP inf.)	kg DCB eq.	0.000471	0.673476	0.031738516	0.641737303	95.29%
	I02+ v2.1 – Aquatic ecotoxicity – Midpoint	kg TEG eq. to water	1.397164	1999.664	94.23704954	1905.427172	95.29%
	ILCD Ecotoxicity freshwater midpoint (v1.09)	CTUe	0.130201	186.3473	8.781886244	177.565456	95.29%
	ReCiPe 2016 Midpoint (E) – Freshwater ecotoxicity	kg 1,4 DB eq.	0.000117	0.167384	0.007888209	0.15949574	95.29%
	TRACI 2.1, Ecotoxicity	CTUe	0.005714	8.177569	0.385379678	7.792189094	95.29%

(*Continued*)

Table 4.1 (Continued)

Impact category	*Life-cycle assessment method*	*Unit*	*Unit environmental impact for 1MJ*	*Reference building environmental impact*	*Green Star environmental impact*	*Reduced environmental impact*	*Reduced percentage*
Eutrophication potential (EP)	CML2001 – Jan. 2016, Eutrophication Potential (EP)	kg PO4 eq.	9.01E-05	1.290E-01	6.078E-03	1.229E-01	95.29%
	I02+ v2.1 – Aquatic eutrophication – Midpoint	kg PO4 eq.	3.91E-07	5.602E-04	2.640E-05	5.338E-04	95.29%
	ILCD Eutrophication freshwater midpoint (v1.09)	kg PO4 eq.	1.29E-07	1.845E-04	8.694E-06	1.758E-04	95.29%
	ReCiPe 2016 Midpoint (H) – Freshwater Eutrophication	kg PO4 eq.	1.29E-07	1.843E-04	8.684E-06	1.756E-04	95.29%
	TRACI 2.1, Eutrophication	kg N eq.	3.58E-05	5.123E-02	2.414E-03	4.881E-02	95.29%
Human toxicity	CML2001 – Jan. 2016, Human Toxicity Potential (HTP inf.)	kg DCB eq.	0.02638	3.776E+01	1.779E+00	3.598E+01	95.29%
	ILCD Human toxicity midpoint, cancer effects (v1.09)	CTUh	1.23E-08	1.753E-05	8.263E-07	1.671E-05	95.29%
	ReCiPe 2016 Midpoint (E) – Human toxicity, cancer	kg 1,4-DB eq.	0.001169	1.673E+00	7.886E-02	1.594E+00	95.29%
	TRACI 2.1, Human toxicity, cancer	CTUh	4.73E-11	6.772E-08	3.191E-09	6.453E-08	95.29%
Photochem. ozone creation potential (POCP)	CML2001 – Jan. 2016, Photochem. Ozone Creation Potential (POCP)	kg C2H4 eq. to air	5.54E-05	0.079	0.004	0.076	95.29%
	I02+ v2.1 – Photochemical oxidation – Midpoint	kg C2H4 eq. to air	8.49E-06	0.012	0.001	0.012	95.29%
	ILCD – Photochemical ozone formation midpoint, human health (v1.09)	kg NMVOC eq.	0.000723	1.035	0.049	0.987	95.29%
	ReCiPe 2016 Midpoint (H) – Photochemical Ozone Formation, Ecosystems	kg NOx eq.	0.525731	752.443	35.460	716.983	95.29%

Ozone layer depletion potential	CML2001 – Jan. 2016 Ozone Layer Depletion Potential (ODP, steady state)	kg R-11 eq.	1.86E-15	2.657E-12	1.252E-13	2.532E-12	95.29%
	I02+ v2.1 – Ozone layer depletion – Midpoint	kg R-11 eq.	1.86E-15	2.657E-12	1.252E-13	2.532E-12	95.29%
	ILCD – Ozone depletion midpoint (v1.09)	kg CFC-11 eq.	2.7E-15	3.863E-12	1.820E-13	3.681E-12	95.29%
	ReCiPe 2016 Midpoint (H) – Stratospheric Ozone Depletion	kg CFC-11 eq.	5.67E-08	8.121E-05	3.827E-06	7.739E-05	95.29%
	TRACI 2.1, Ozone Depletion Air	kg CFC-11 eq.	1.86E-15	2.659E-12	1.253E-13	2.534E-12	95.29%
Terrestrial acidification	CML2001 – Jan. 2016, Acidification Potential (AP)	kg SO_2 eq.	0.001049	1.502	0.071	1.431	95.29%
	I02+ v2.1 – Terrestrial acidification/nitrification – Midpoint	kg SO_2 eq. to air	0.004244	6.074	0.286	5.788	95.29%
	ILCD Acidification midpoint (v1.09)	Mole of H^+ eq.	0.001268	1.815	0.086	1.730	95.29%
	ReCiPe 2016 Midpoint (E) – Terrestrial Acidification	kg SO_2 eq.	0.000832	1.191	0.056	1.135	95.29%
	TRACI 2.1, Acidification	kg SO2 eq.	0.001065	1.525	0.072	1.453	95.29%
Terrestrial ecotoxicity	CML2001 – Jan. 2016, Terrestrial Ecotoxicity Potential (TETP inf.)	kg DCB eq.	0.00039	0.558	0.026	0.532	95.29%
	ReCiPe 2016 Midpoint (H) – Terrestrial ecotoxicity	kg 1,4-DB eq.	0.12171	174.195	8.209	165.986	95.29%

(Continued)

Table 4.2 Summary of life-cycle impacts of 1 m^3 timber

Impact category	*Life-cycle assessment method*	*Unit*	*Original environmental impact*	*Green Star environmental impact*	*Reduced environmental impact*
Global Warming Potential (GWP 100 years)	CML2001 – Jan. 2016, Global Warming Potential (GWP 100 years)	kg CO_2 eq.	0.292252	0.014613	0.27764
	ILCD – Climate change midpoint, incl. biogenic carbon (v1.09)	kg CO_2 eq.	0.293118	0.014656	0.278462
	ReCiPe 2016 Midpoint (H) – Climate change, incl. biogenic carbon	kg CO_2 eq.	0.279487	0.013974	0.265513
	TRACI 2.1, Global Warming Air, incl. biogenic carbon	kg CO_2 eq.	0.291444	0.014572	0.276872
Aquatic Ecotoxicity	CML2001 – Jan. 2016, Freshwater Aquatic Ecotoxicity Pot. (FAETP inf.)	kg DCB eq.	0.001325	6.62E-05	0.001258
	I02+ v2.1 – Aquatic ecotoxicity – Midpoint	kg TEG eq. to water	6.600667	0.330033	6.270634
	ILCD Ecotoxicity freshwater midpoint (v1.09)	CTUe	0.030885	0.001544	0.029341
	ReCiPe 2016 Midpoint (E) – Freshwater ecotoxicity	kg 1,4 DB eq.	4.53E-05	2.27E-06	4.3E-05
	TRACI 2.1, Ecotoxicity	CTUe	0.015231	0.000762	0.01447
Eutrophication Potential (EP),	CML2001 – Jan. 2016, Eutrophication Potential (EP)	kg PO4 eq.	0.000293	1.46E-05	0.000278
	I02+ v2.1 – Aquatic eutrophication – Midpoint	kg PO4 eq.	3.55E-06	1.78E-07	3.38E-06
	ILCD Eutrophication freshwater midpoint (v1.09)	kg PO4 eq.	1.17E-06	5.85E-08	1.11E-06
	ReCiPe 2016 Midpoint (H) – Freshwater Eutrophication	kg PO4 eq.	1.17E-06	5.84E-08	1.11E-06
	TRACI 2.1, Eutrophication	kg N eq.	0.000134	6.69E-06	0.000127
Human toxicity	CML2001 – Jan. 2016, Human Toxicity Potential (HTP inf.)	kg DCB eq.	0.01989	0.000994	0.018895
	ILCD Human toxicity midpoint, cancer effects (v1.09)	CTUh	1.17E-09	5.86E-11	1.11E-09
	ReCiPe 2016 Midpoint (E) – Human toxicity, cancer	kg 1,4-DB eq.	0.00435	0.000218	0.004133
	TRACI 2.1, Human toxicity, cancer	CTUh	2.23E-10	1.12E-11	2.12E-10

Photochem. Ozone Creation Potential (POCP)	CML2001 – Jan. 2016, Photochem. Ozone Creation Potential (POCP)	kg C_2H_4 eq. to air	0.000106	5.3E-06	0.000101
	I02+ v2.1 – Photochemical oxidation – Midpoint	kg C_2H_4 eq. to air	0.000333	1.67E-05	0.000317
	ILCD – Photochemical ozone formation midpoint, human health (v1.09)	kg NMVOC eq.	0.002531	0.000127	0.002404
	ReCiPe 2016 Midpoint (H) – Photochemical Ozone Formation, Ecosystems	kg NOx eq.	0.704549	0.035227	0.669321
Ozone Layer Depletion Potential	CML2001 – Jan. 2016 Ozone Layer Depletion Potential (ODP, steady state)	kg R-11 eq.	5.84E-13	2.92E-14	5.55E-13
	I02+ v2.1 – Ozone layer depletion – Midpoint	kg R-11 eq.	5.84E-13	2.92E-14	5.55E-13
	ILCD – Ozone depletion midpoint (v1.09)	kg CFC-11 eq.	5.87E-13	2.93E-14	5.57E-13
	ReCiPe 2016 Midpoint (H) – Stratospheric Ozone Depletion	kg CFC-11 eq.	2.55E-07	1.27E-08	2.42E-07
	TRACI 2.1, Ozone Depletion Air	kg CFC-11 eq.	5.84E-13	2.92E-14	5.55E-13
Terrestrial Acidification	CML2001 – Jan. 2016, Acidification Potential (AP)	kg SO_2 eq.	0.001539	7.7E-05	0.001462
	I02+ v2.1 – Terrestrial acidification/nitrification – Midpoint	kg SO_2 eq. to air	0.000333	1.67E-05	0.000317
	ILCD Acidification midpoint (v1.09)	Mole of H^+ eq.	0.002034	0.000102	0.001932
	ReCiPe 2016 Midpoint (E) – Terrestrial Acidification	kg SO_2 eq.	0.00113	5.65E-05	0.001073
	TRACI 2.1, Acidification	kg SO_2 eq.	0.001889	9.45E-05	0.001795
Terrestrial ecotoxicity	CML2001 – Jan. 2016, Terrestric Ecotoxicity Potential (TETP inf.)	kg DCB eq.	0.000415	2.08E-05	0.000394
	I02+ v2.1 – Terrestrial ecotoxicity – Midpoint	kg TEG eq. to soil	0.194739	0.009737	0.185002
	ReCiPe 2016 Midpoint (H) – Terrestrial ecotoxicity	kg 1,4-DB eq.	0.074684	0.003734	0.07095

Table 4.3 Summary of life-cycle impact of 1 kg steel (during 50 years of an office building's lifetime)

Impact category	*Life-cycle assessment method*	*Unit*	*Original environmental impact*	*Green Star environmental impacts*	*Reduced environmental impacts*
Global Warming Potential (GWP 100 years)	CML2001 – Jan. 2016, Global Warming Potential (GWP 100 years)	kg CO_2 eq.	0.036335	0.034518	0.001817
	I02+ v2.1 – Global warming 500yr – Midpoint	kg CO_2 eq.	0.035941	0.034144	0.001797
	ILCD – Climate change midpoint, incl. biogenic carbon (v1.09)	kg CO_2 eq.	0.036273	0.034459	0.001814
	ReCiPe 2016 Midpoint (H) – Climate change, incl. biogenic carbon	kg CO_2 eq.	0.035344	0.033577	0.001767
	TRACI 2.1, Global Warming Air, incl. biogenic carbon	kg CO_2 eq.	0.036245	0.034433	0.001812
Aquatic Ecotoxicity	CML2001 – Jan. 2016, Freshwater Aquatic Ecotoxicity Pot. (FAETP inf.)	kg DCB eq.	4.73E-05	4.49E-05	2.36E-06
	I02+ v2.1 – Aquatic ecotoxicity – Midpoint	kg TEG eq. to water	0.180052	0.17105	0.009003
	ILCD Ecotoxicity freshwater midpoint (v1.09)	CTUe	0.001794	0.001704	8.97E-05
	ReCiPe 2016 Midpoint (E) – Freshwater ecotoxicity	kg 1,4 DB eq.	3.98E-06	3.78E-06	1.99E-07
	TRACI 2.1, Ecotoxicity	CTUe	0.000923	0.000877	4.62E-05
Eutrophication Potential (EP),	CML2001 – Jan. 2016, Eutrophication Potential (EP)	kg PO_4 eq.	7.86E-06	7.46E-06	3.93E-07
	I02+ v2.1 – Aquatic eutrophication – Midpoint	kg PO_4 eq.	2E-07	1.9E-07	1E-08
	ILCD Eutrophication freshwater midpoint (v1.09)	kg PO_4 eq.	6.58E-08	6.25E-08	3.29E-09
	ReCiPe 2016 Midpoint (H) – Freshwater Eutrophication	kg PO_4 eq.	6.57E-08	6.24E-08	3.29E-09
	TRACI 2.1, Eutrophication	kg N eq.	4.79E-06	4.55E-06	2.4E-07
Human toxicity	CML2001 – Jan. 2016, Human Toxicity Potential (HTP inf.)	kg DCB eq.	0.002741	0.002604	0.000137
	ILCD Human toxicity midpoint, cancer effects (v1.09)	CTUh	5.73E-11	5.44E-11	2.87E-12
	ReCiPe 2016 Midpoint (E) – Human toxicity, cancer	kg 1,4-DB eq.	0.000139	0.000132	6.94E-06
	TRACI 2.1, Human toxicity, cancer	CTUh	1.57E-11	1.49E-11	7.83E-13

Photochem. Ozone Creation Potential (POCP)	CML2001 – Jan. 2016, Photochem. Ozone Creation Potential (POCP)	kg C_2H_4 eq. to air	1.17E-05	1.11E-05	5.86E-07
	I02+ v2.1 – Photochemical oxidation – Midpoint	kg C_2H_4 eq. to air	2.06E-06	1.95E-06	1.03E-07
	ILCD – Photochemical ozone formation midpoint, human health (v1.09)	kg NMVOC eq.	6.73E-05	6.39E-05	3.37E-06
	ReCiPe 2016 Midpoint (H) – Photochemical Ozone Formation, Ecosystems	kg NOx eq.	0.023327	0.022161	0.001166
Ozone Layer Depletion Potential	CML2001 – Jan. 2016 Ozone Layer Depletion Potential (ODP, steady state)	kg R-11 eq.	1.11E-14	1.05E-14	5.53E-16
	I02+ v2.1 – Ozone layer depletion – Midpoint	kg R-11 eq.	1.61E-14	1.53E-14	8.04E-16
	ILCD – Ozone depletion midpoint (v1.09)	kg CFC-11 eq.	1.63E-14	1.55E-14	8.16E-16
	ReCiPe 2016 Midpoint (H) – Stratospheric Ozone Depletion	kg CFC-11 eq.	7.65E-09	7.27E-09	3.83E-10
	TRACI 2.1, Ozone Depletion Air	kg CFC-11 eq.	1.61E-14	1.53E-14	8.04E-16
Terrestrial Acidification	CML2001 – Jan. 2016, Acidification Potential (AP)	kg SO_2 eq.	7.45E-05	7.08E-05	3.72E-06
	I02+ v2.1 – Terrestrial acidification/nitrification – Midpoint	kg SO_2 eq. to air	0.000313	0.000297	1.56E-05
	ILCD Acidification midpoint (v1.09)	Mole of H^+ eq.	8.87E-05	8.43E-05	4.43E-06
	ReCiPe 2016 Midpoint (E) – Terrestrial Acidification	kg SO_2 eq.	5.76E-05	5.47E-05	2.88E-06
	TRACI 2.1, Acidification	kg SO_2 eq.	7.71E-05	7.33E-05	3.86E-06
Terrestrial ecotoxicity	CML2001 – Jan. 2016, Terrestric Ecotoxicity Potential (TETP inf.)	kg DCB eq.	6.5E-05	6.18E-05	3.25E-06
	ReCiPe 2016 Midpoint (H) – Terrestrial ecotoxicity	kg 1,4-DB eq.	7.4E-07	7.03E-07	3.7E-08

By applying these criteria requirements of Credit 19, the project will successfully achieve 2 points to add to the total. The minimum grade of hot rolled structural reinforcement bar in this category is 350 MPa. For cold-formed section, the minimum grade is 450 MPa, and the value for the welded section is 400 MPa (Green Building Council of Australia, 2015a).

4.3.2 *Environmental impact analysis for Green Star concrete alternatives*

Table 4.4 summarizes the results of greenhouse gas emissions for 100% ordinary Portland cement and non-air-entrained concrete for different aggregate types with 25 mm aggregate size as well as the required slump range within 30 to 50 mm for concrete strength of 20–100 MPa. As seen in this table, although greenhouse gas emission results are very much different among options, 0.5 points of Credit 19 is achieved merely by reclaiming water or using alternative fine aggregates.

The selection of the appropriate aggregate is crucial for all structural concrete and is independent of concrete strength (Caldarone, 2009). Changing aggregate types affects concrete sustainability (Lemay et al., 2013). When using the proposed model, the amount of greenhouse gas emissions generated from angular coarse aggregate is higher than that of concrete using the remaining three aggregate types, namely, sub-angular aggregate, gravel with some crushed particles, and rounded gravel.

For example, greenhouse gas emissions of angular coarse concrete, sub-angular aggregate concrete, gravel with some crushed particles concrete, and rounded gravel concrete for the 32 MPa concrete are 0.1280, 0.1287, 0.1293, and 0.1298 kg CO_2-eq/m^3, respectively, as is seen in Table 4.4. This is because the cement amount used in angular-coarse-concrete is higher than in concrete obtained from using the remaining three aggregate types causing higher greenhouse gas emissions. For a given water-to-cement ratio, compared with other aggregate types, rounded aggregate demands lower cement content because of its smaller surface area for a given volume, which is termed the "specific surface area" (Breins Engineering, 2016).

Rough-textured and angular-coarse aggregate provide stronger mechanical bonding and are generally more appropriate for high-strength concrete than smooth-textured aggregate (Neville, 1997). For concrete produced using identical material and alike proportions, crushed-coarse aggregate from fine-grained diabase and limestone likely yields the highest concrete strength (Caldarone, 2009). Hence, this type of aggregate should be considered as "green" and be chosen as a preferred aggregate type for concrete production.

Table 4.5 and Figure 4.1 show that the results of greenhouse gas emissions change depending on the percentage of supplementary cement materials as well as the strength of concrete. The higher the usage of cementitious materials, the lower the level of greenhouse gas emissions released during the life-cycle of concrete. The estimated data also reveal that greenhouse gas emissions of concrete have a non-linear relationship with concrete strength. For instance, greenhouse gas emissions released from 100 MPa concrete with gravel of maximum 25 mm

Table 4.4 Greenhouse gas emissions results by using coarse aggregate, alternative fine aggregate, and water reclaiming, kg CO_2-eq/m^3

Type of aggregate	*(a)*	*(b)*	*(c)*	*Achieved Green Star Credit 19 points*	*Concrete strength grade (following AS1379)*							
					20	*25*	*32*	*40*	*50*	*65*	*80*	*100*
Angular coarse (crushed stone)				0	0.1245	0.1252	0.1280	0.1509	0.1589	0.1810	0.2019	0.2157
	X			0.5	0.1246	0.1246	0.1281	0.1510	0.1590	0.1811	0.2020	0.2158
		X		0.5	0.1236	0.1236	0.1270	0.1498	0.1576	0.1795	0.2003	0.2140
			X	0.5	0.1228	0.1228	0.1228	0.1230	0.1045	0.1047	0.1049	0.1050
Sub-angular aggregate				0	0.1253	0.1258	0.1287	0.1517	0.1599	0.1821	0.2031	0.2170
	X			0.5	0.1254	0.1252	0.1288	0.1518	0.1599	0.1822	0.2032	0.2171
		X		0.5	0.1244	0.1242	0.1278	0.1506	0.1585	0.1806	0.2015	0.2153
			X	0.5	0.1235	0.1234	0.1235	0.1237	0.1051	0.1053	0.1055	0.1056
Gravel with some crushed particles				0	0.1261	0.1266	0.1293	0.1524	0.1609	0.1833	0.2045	0.2184
	X			0.5	0.1262	0.1260	0.1294	0.1525	0.1610	0.1834	0.2046	0.2185
		X		0.5	0.1252	0.1249	0.1284	0.1514	0.1596	0.1818	0.2028	0.2167
			X	0.5	0.1243	0.1241	0.1242	0.1243	0.1058	0.1060	0.1062	0.1063
Rounded gravel				0	0.1261	0.1270	0.1298	0.1529	0.1614	0.1838	0.2051	0.2191
	X			0.5	0.1262	0.1264	0.1299	0.1530	0.1615	0.1839	0.2051	0.2192
		X		0.5	0.1252	0.1254	0.1288	0.1518	0.1600	0.1823	0.2034	0.2173
			X	0.5	0.1243	0.1246	0.1246	0.1246	0.1061	0.1063	0.1065	0.1066

Note: (a): 50% water reclaimed, (b): Crushed slag aggregate or other alternative materials, (c): Manufactured sand or other alternative materials.

Table 4.5 Greenhouse gas emissions and required cement for concrete using gravel sizes 25 mm, 30 mm to 50 mm slump, kg CO_2-eq/m^3

Percentage of cementitious material			*Concrete strength grade (following AS1379), no alternative fine aggregate, no reclaimed water*							
			20	25	32	40	50	65	80	100
100% OPC		(a)	0.1245	0.1252	0.1280	0.1509	0.1589	0.1810	0.2019	0.2157
		(b)	335	337	345	412	440	506	569	611
30% SCM	SCM1	(a)	0.0989	0.0989	0.1016	0.1195	0.1259	0.1431	0.1595	0.1703
	SCM2	(a)	0.1018	0.1018	0.1046	0.1230	0.1296	0.1474	0.1642	0.1753
	SCM3	(a)	0.0901	0.0901	0.0926	0.1088	0.1146	0.1301	0.1449	0.1547
		(b)	234	236	241	288	308	354	398	428
40% SCM	SCM1	(a)	0.0783	0.0783	0.0804	0.0942	0.0993	0.1127	0.1254	0.1337
	SCM2	(a)	0.0941	0.0941	0.0967	0.1136	0.1197	0.1361	0.1515	0.1618
	SCM3	(a)	0.0786	0.0786	0.0807	0.0946	0.0997	0.1131	0.1258	0.1342
		(b)	201	202	207	247	264	304	341	367

Note: (a): Greenhouse gas emissions, (b): Required cement amount, SCM1: Fly ash; SCM2: Silica fume; SCM3: Blast furnace slag

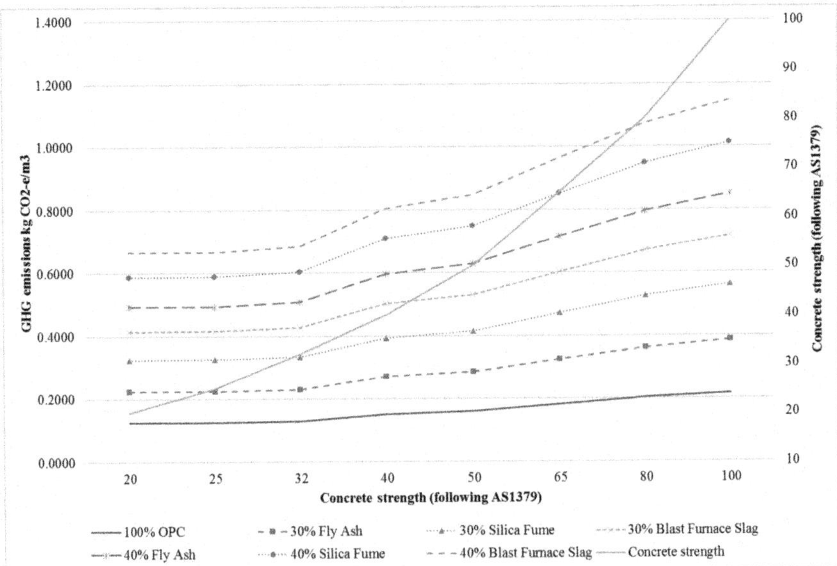

Figure 4.1 Greenhouse gas emissions for concrete options in Table 4.5

aggregate size are 0.2157 kg CO_2-eq/m^3, approximately 1.73 times more than that of the 20 MPa concrete with the same aggregate size (0.1245 kg CO_2-eq/m^3).

In terms of material composition for concrete production, the amount of greenhouse gas increased consistently with the increasing of the amount of required cement used for the concrete, as shown in Figure 4.2. The amount of greenhouse gas produced in 1 m^3 concrete stems predominantly from cement. The total amount of greenhouse gas emissions of the 20 MPa concrete is 0.1245 kg

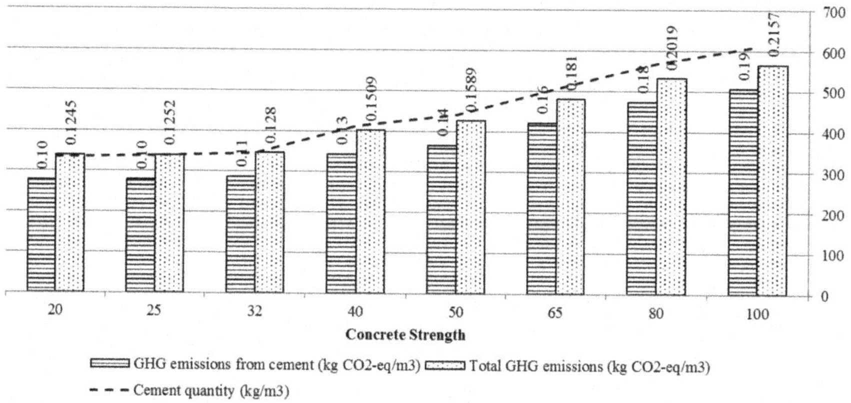

Figure 4.2 Relationship between cement and greenhouse gas emissions

CO_2-eq/m^3, comprising 0.1 kg CO_2-eq/m^3 greenhouse gas emissions from cement. In another case of 100 MPa concrete, the greenhouse gas emission from cement is 0.2157 kg CO_2-eq/m^3, while the total greenhouse gas emissions from 1 m^3 of this type of concrete is 0.19 kg CO_2-eq/m^3.

4.3.3 *Concrete replacement of ordinary Portland cement with supplementary cement materials*

The concrete mixture proportion and greenhouse gas emissions can be automatically estimated when one or a combination of more than one variable are changed: the concrete strength grade, aggregate type or maximum sizes of aggregate, and exposure environment conditions (Standards Association of Australia, 2007).

Table 4.5 summarizes the model's results in achieving the point of Credit 19 for 20 MPa and 100 MPa concrete using gravel with some crushed particles. This table demonstrates the reduction of life-cycle greenhouse gas emissions by using supplementary cement materials in concrete.

Due to the lowest amount of life-cycle greenhouse gas emissions, silica fume is found to be the optimal supplementary cement material that can be used to reduce greenhouse gas emissions among the three studied supplementary cement materials (Table 4.15). The 100 MPa concrete produces 0.2158 kg CO_2 -eq/m^3 greenhouse gas emissions from 100% OPC concrete, while concrete with 30% silica fume replacement generates 0.1753 kg CO_2-eq/m^3 greenhouse gas emissions (Table 4.6).

Fly ash and blast furnace slag concrete with the same strengths have a slightly higher quantity of greenhouse gas emissions at 0.1337 and 0.1342 kg CO_2-eq/m^3, respectively (Table 4.5 and Figure 4.1). According to the model's result, independent of the percentage of supplementary cement materials, greenhouse gas emissions of silica fume concrete are always less than those from the remaining two

Table 4.6 Greenhouse gas emissions by varying Credit 19 options for 100 MPa concrete using gravel and crushed particles

Mixture	0–0.5W	0–0.5Alt(1)	0–0.5Alt(2)	1SCM1–0	1SCM2–0	1SCM3–0
Credit points	0.5	0.5	0.5	1	1	1
GHG	0.2158	0.2140	0.2156	0.1703	0.1753	0.1547
Mixture	1SCM1–0.5Alt(1)	1SCM2–0.5Alt(1)	1SCM3–0.5Alt(1)	1SCM1–0.5Alt(2)	1SCM2–0.5Alt(2)	1SCM3–0.5Alt(2)
Credit points	1.5	1.5	1.5	1.5	1.5	1.5
GHG	0.1682	0.1733	0.1527	0.1696	0.1748	0.1541
Mixture	1SCM1–1Alt(1)	1SCM2–1Alt(1)	1SCM3–1Alt(1)	1SCM1–1Alt(2)	1SCM2–1Alt(2)	1SCM3–1Alt(2)
Credit points	2	2	2	2	2	2
GHG	0.1680	0.1732	0.1525	0.1694	0.1746	0.1539
Mixture	2SCM1–0.5Alt(1)	2SCM2–0.5Alt(1)	2SCM3–0.5Alt(1)	2SCM1–0.5Alt(2)	2SCM2–0.5Alt(2)	2SCM3–0.5Alt(2)
Credit points	2.5	2.5	2.5	2.5	2.5	2.5
GHG	0.1529	0.1598	0.1322	0.1543	0.1612	0.1336
Mixture	2SCM1–1Alt(1)	2SCM2–1Alt(1)	2SCM3–1Alt(1)	2SCM1–1Alt(2)	2SCM2–1Alt(2)	2SCM3–1Alt(2)
Credit points	3	3	3	3	3	3
GHG	0.1527	0.1596	0.1321	0.1541	0.1610	0.1335

Note: (a): 50% water reclaiming; (b): Alt1: Crushed slag aggregate or other alternative materials, Alt2: Manufactured sand or other alternative materials; SCM1: Fly ash; SCM2: Silica fume; SCM3: Blast furnace slag

supplementary cement materials deploying concrete types for both desired point levels under Credit 19 (Tam et al., 2019).

Although these options achieve two credit points for Credit 19, greenhouse gas emission results with supplementary cement materials combinations using reclaiming water or alternative fine aggregate are higher than using only supplementary cement materials. Moreover, adjusting the water-to-cement ratio and using alternative fine aggregates are two effective methods in achieving the credit points while their greenhouse gas emissions remain low (Table 4.4). Based on this analysis, high-strength concrete was found to be more sustainable than regular strength concrete (Table 4.4 and Table 4.5) (Aïtcin & Mindess, 2011). The water-to-binder ratio plays an active role in managing the concrete strength but also in controlling the concrete's life-cycle greenhouse gas emissions. The model shows that greenhouse gas emissions are increased about 1.73 times for five-time concrete-strength increase from 0.1245 kg CO_2 -eq/m^3 (20 MPa) to 0.2157 kg CO_2-eq/m^3 (100 MPa) (Table 4.7).

This means that it is crucial to lower the water-to-binder ratio in concrete, and this ratio should be carefully monitored as precisely as possible, such that it can be used to build a durable and sustainable concrete structure (Aïtcin & Mindess, 2011). The trade-off between life-cycle greenhouse gas emissions and concrete

Table 4.7 Water-to-binder ratio versus greenhouse gas emissions

Percentage of cementitious material		*Concrete strength grade (following AS1379), no alternative fine aggregate, no reclaimed water*							
		20	25	32	40	50	65	80	100
100%	(a)	0.1245	0.1252	0.1280	0.1509	0.1589	0.1810	0.2019	0.2157
OPC	(b)	0.430	0.427	0.417	0.350	0.280	0.243	0.216	0.201

Note: (a): greenhouse gas emissions (kg CO2-eq/m^3), (b): Water-to-binder ratio.

strength is also clear, and this compromise should be carefully studied, so that concrete with a specific strength and life-cycle greenhouse gas emissions can be optimally designed. Choosing the coarse-aggregate type for concrete strength improvement, or producing the "green" concrete, depends on particle types and shapes. The rough-coarse and angular-coarse aggregate types produce stronger concrete, but at the same time produce higher greenhouse gas emissions because of the higher cement content. It is thus apparent that a trade-off between aggregate type and life-cycle greenhouse gas emissions also exists. These relationships can be further optimized when using the proposed model, such that the most suitable concrete can be designed under specific conditions and constraints set by companies and designers, which is the goal of the proposed model in this book.

Additional verification processes using different mix designs, gravel sizes, and concrete strengths have been conducted, and matching findings have been obtained. The size of aggregates that are used to design the concrete mixture plays a significant role in managing the life-cycle greenhouse gas emissions of concrete. The model is verified continuously using different datasets and materials in different conditions to minimize its computation errors. This process is understandably lengthy and quite intensive. Within the scope of this book, verification for greenhouse gas emissions analysis has been successfully achieved. By changing the ratio of cement, aggregate, and water, this book helps designers understand and achieve the credit points in Green Star. The environmental impacts of 1 m^3 of each type of concrete are extracted from GaBi 8.7 and presented in Table 4.8 to Table 4.15.

4.4 Summary

This chapter presents the life-cycle greenhouse gas emissions analysis sequence and results of the building's major materials and choice of envelope. According to the estimation process in Chapter 3, designers and quantity surveyors can first select the building fabric layers, which are determined in the handbook of the Insulation Council of Australia and New Zealand (2014) for the desired and the reference building. Then, the amount of energy consumption needed to balance the indoor and outdoor temperatures is calculated. The reduction percentage of energy consumption and greenhouse gas emissions between the desired building and the reference building referred to in Table 3.7 will help the project achieve awarded points in Green Star Credit 15.

Table 4.8 Life-cycle greenhouse gas emissions of 1 kg 20 MPa concrete

Impact category	Life-cycle assessment method / Concrete mixture option	Unit	0-0-0 (1)0-0	0-1-0 (2)0-0.5W	0-0-1 (3)0-0.5Cs	0-0-2 (4)0-0.5S	1-0-0 (5)1F-0	2-0-0 (7)1S-0	3-0-0 (6)1B-0	1-1-0 (8)1F-0.5W	2-1-0 (10)1S-0.5W	3-1-0 (9)1B-0.5W	1-0-1 (11)1F-0.5Cs	2-0-1 (13)1S-0.5Cs	3-0-1 (12)1B-0.5Cs	1-0-2 (14)1F-0.5S	2-0-2 (16)1S-0.5S	3-0-2 (15)1B-0.5S
	Points to achieve			0.5	0.5	0.5	1	1	1	1.5	1.5	1.5	1.5	1.5	1.5	1.5	1.5	1.5
Global Warming Potential (GWP 100 years)	CML2001 - Jan. 2016, Global Warming Potential (GWP 100 years)	kg CO_2 eq.	0.1245	0.1246	0.1236	0.1228	0.0989	0.1018	0.0901	0.0986	0.1015	0.0899	0.0976	0.1005	0.0889	0.0982	0.1011	0.0895
	IO2+ v2.1 - Global warming 500yr - Midpoint	kg CO_2 eq.	0.1241	0.1241	0.1229	0.1225	0.0977	0.1005	0.0895	0.0977	0.1004	0.0894	0.0965	0.0993	0.0886	0.0971	0.0999	0.0892
	ILCD - Climate change midpoint, incl biogenic carbon (v1.09)	kg CO_2 eq.	0.1244	0.1244	0.1232	0.1227	0.0984	0.1011	0.0898	0.0984	0.1011	0.0897	0.0972	0.0999	0.0885	0.0978	0.1005	0.0891
	ReCiPe 2016 Midpoint (H) - Climate change, incl biogenic carbon	kg CO_2 eq.	0.1232	0.1232	0.1221	0.1216	0.0882	0.0997	0.0888	0.0970	0.0997	0.0888	0.0958	0.0986	0.0876	0.0964	0.0991	0.0882
	TRACI 2.1, Global Warming Air, incl. biogenic carbon	kg CO_2 eq.	0.1244	0.1243	0.1232	0.1227	0.0891	0.1011	0.0897	0.0983	0.1010	0.0897	0.0972	0.0999	0.0885	0.0978	0.1005	0.0891
Aquatic Ecotoxicity	CML2001 - Jan. 2016, Freshwater Aquatic Ecotoxicity Pot. (FAETP inf.)	kg DCB eq.	1.019E-04	1.014E-04	9.847E-05	9.462E-05	8.565E-05	9.027E-05	7.712E-05	8.503E-05	8.974E-05	7.659E-05	8.211E-05	8.683E-05	7.368E-05	8.394E-05	8.869E-05	7.551E-05
	IO2+ v2.1 - Aquatic ecotoxicity - Midpoint	kg TEG eq. to water	2.674E-01	2.650E-01	2.508E-01	2.259E-01	2.758E-01	3.113E-01	2.170E-01	2.734E-01	3.089E-01	2.146E-01	2.592E-01	2.947E-01	2.004E-01	2.691E-01	3.047E-01	2.103E-01
	ILCD Ecotoxicity freshwater midpoint (v1.09)	CTUe	2.222E-03	2.199E-03	2.096E-03	1.975E-03	2.078E-03	1.844E-03	1.768E-03	2.055E-03	1.821E-03	1.745E-03	1.952E-03	1.718E-03	1.642E-03	2.022E-03	1.789E-03	1.711E-03
	ReCiPe 2016 Midpoint (E) - Freshwater ecotoxicity	kg 1,4 DB eq.	4.146E-06	4.104E-06	3.985E-06	3.775E-06	3.174E-06	3.842E-06	3.197E-06	3.670E-06	3.799E-06	3.155E-06	3.551E-06	3.680E-06	3.036E-06	3.636E-06	3.767E-06	3.120E-06
	TRACI 2.1, Ecotoxicity	CTUe	1.251E-03	1.239E-03	1.187E-03	1.097E-03	9.727E-04	9.664E-04	9.793E-04	1.211E-03	9.545E-04	9.674E-04	1.159E-03	9.026E-04	9.155E-04	1.198E-03	9.423E-04	9.546E-04
Eutrophication Potential (EP)	CML2001 - Jan. 2016, Eutrophication Potential (EP)	kg PO_4 eq.	2.398E-05	2.411E-05	2.309E-05	2.225E-05	2.163E-05	2.236E-05	1.947E-05	2.141E-05	2.215E-05	1.926E-05	2.039E-05	2.112E-05	1.823E-05	2.101E-05	2.174E-05	1.885E-05
	IO2+ v2.1 - Aquatic eutrophication - Midpoint	kg PO_4 eq.	2.358E-07	2.265E-07	2.223E-07	1.788E-07	1.944E-07	3.168E-07	1.916E-07	1.851E-07	3.075E-07	1.829E-07	1.809E-07	3.094E-07	1.781E-07	1.885E-07	3.110E-07	1.858E-07
	ILCD Eutrophication freshwater midpoint (v1.09)	kg PO_4 eq.	7.559E-08	7.447E-08	7.309E-08	5.883E-08	6.386E-08	1.042E-07	6.295E-08	6.082E-08	1.011E-07	5.992E-08	5.944E-08	9.977E-08	5.853E-08	6.194E-08	1.023E-07	6.104E-08
	ReCiPe 2016 Midpoint (H) - Freshwater Eutrophication	kg PO_4 eq.	7.744E-08	7.440E-08	7.302E-08	5.877E-08	6.286E-08	1.041E-07	6.290E-08	6.076E-08	1.011E-07	5.986E-08	5.938E-08	9.968E-08	5.848E-08	6.188E-08	1.022E-07	6.099E-08
	TRACI 2.1, Eutrophication	kg N eq.	1.233E-05	1.228E-05	1.182E-05	1.143E-05	9.915E-06	1.261E-05	1.014E-05	1.162E-05	1.256E-05	1.009E-05	1.116E-05	1.210E-05	9.630E-06	1.142E-05	1.236E-05	9.895E-06
Human toxicity	CML2001 - Jan. 2016, Human Toxicity Potential (HTP inf.)	kg DCB eq.	2.686E-02	2.685E-02	2.681E-02	2.674E-02	1.917E-02	1.942E-02	1.888E-02	1.917E-02	1.942E-02	1.887E-02	1.913E-02	1.938E-02	1.883E-02	1.915E-02	1.940E-02	1.885E-02
	ILCD Human toxicity midpoint, cancer effects (v1.09)	CTUh	9.946E-11	9.824E-11	9.280E-11	8.690E-11	8.949E-11	8.874E-11	8.193E-11	8.827E-11	8.752E-11	8.071E-11	8.233E-11	8.158E-11	7.477E-11	8.588E-11	8.524E-11	7.832E-11
	ReCiPe 2016 Midpoint (E) - Human toxicity, cancer	kg 1,4-DB eq.	3.039E-04	2.929E-04	2.804E-04	2.393E-04	2.556E-04	2.946E-04	2.591E-04	2.748E-04	2.836E-04	2.480E-04	2.623E-04	2.711E-04	2.355E-04	2.738E-04	2.829E-04	2.471E-04
	TRACI 2.1, Human toxicity, cancer	CTUh	4.168E-11	4.148E-11	4.030E-11	3.739E-11	3.130E-11	3.423E-11	3.133E-11	3.297E-11	3.402E-11	3.113E-11	3.179E-11	3.284E-11	2.994E-11	3.250E-11	3.356E-11	3.066E-11
Photochem. Ozone Creation Potential (POCP)	CML2001 - Jan. 2016, Photochem. Ozone Creation Potential (POCP)	kg C_2H_4 eq. to air	1.318E-05	1.316E-05	1.430E-05	1.323E-05	7.742E-06	9.067E-06	7.393E-06	7.722E-06	9.046E-06	7.372E-06	8.865E-06	1.019E-05	8.515E-06	8.310E-06	9.636E-06	7.961E-06
	IO2+ v2.1 - Photochemical oxidation - Midpoint	kg C_2H_4 eq. to air	6.896E-06	6.879E-06	6.632E-06	6.273E-06	5.842E-06	5.997E-06	5.287E-06	5.825E-06	5.980E-06	5.270E-06	5.578E-06	5.733E-06	5.023E-06	5.708E-06	5.865E-06	5.154E-06
	ILCD - Photochemical ozone formation midpoint, human health (v1.09)	kg NMVOC eq.	1.903E-04	1.902E-04	1.825E-04	1.782E-04	1.425E-04	1.607E-04	1.461E-04	1.585E-04	1.605E-04	1.459E-04	1.509E-04	1.529E-04	1.383E-04	1.548E-04	1.567E-04	1.421E-04
	ReCiPe 2016 Midpoint (H) - Photochemical Ozone Formation, Ecosystems	kg NO_x eq.	2.552E-02	2.543E-02	2.498E-02	2.382E-02	1.923E-02	3.835E-02	1.925E-02	2.944E-02	3.826E-02	1.916E-02	2.899E-02	3.781E-02	1.871E-02	2.926E-02	3.808E-02	1.898E-02
Ozone Layer Depletion Potential	CML2001 - Jan. 2016 Ozone Layer Depletion Potential (ODP, steady state)	kg R-11 eq.	3.628E-13	3.627E-13	3.621E-13	3.606E-13	3.584E-13	3.717E-13	3.583E-13	3.583E-13	3.716E-13	3.582E-13	3.578E-13	3.711E-13	3.576E-13	3.581E-13	3.714E-13	3.580E-13
	IO2+ v2.1 - Ozone layer depletion - Midpoint	kg R-11 eq.	3.628E-13	3.627E-13	3.621E-13	3.606E-13	3.584E-13	3.717E-13	3.583E-13	3.583E-13	3.716E-13	3.582E-13	3.578E-13	3.711E-13	3.576E-13	3.581E-13	3.714E-13	3.580E-13
	ILCD - Ozone depletion midpoint (v1.09)	kg CFC-11 eq.	3.630E-13	3.629E-13	3.624E-13	3.609E-13	3.586E-13	3.721E-13	3.585E-13	3.585E-13	3.720E-13	3.584E-13	3.580E-13	3.715E-13	3.578E-13	3.583E-13	3.718E-13	3.581E-13
	ReCiPe 2016 Midpoint (H) - Stratospheric Ozone Depletion	kg CFC-11 eq.	1.776E-08	1.778E-08	1.728E-08	1.693E-08	1.400E-08	2.189E-08	1.423E-08	2.041E-08	2.178E-08	1.412E-08	2.004E-08	2.141E-08	1.375E-08	2.030E-08	2.167E-08	1.400E-08
	TRACI 2.1, Ozone Depletion Air	kg CFC-11 eq.	3.847E-13	3.846E-13	3.840E-13	3.826E-13	3.802E-13	3.937E-13	3.802E-13	3.297E-11	3.936E-13	3.801E-13	3.797E-13	3.930E-13	3.796E-13	3.800E-13	3.933E-13	3.799E-13
Terrestrial Acidification	CML2001 - Jan. 2016, Acidification Potential (AP)	kg SO_2 eq.	1.845E-04	1.837E-04	1.790E-04	1.747E-04	1.542E-04	1.559E-04	1.392E-04	1.533E-04	1.551E-04	1.384E-04	1.486E-04	1.504E-04	1.337E-04	1.514E-04	1.531E-04	1.364E-04
	IO2+ v2.1 - Terrestrial acidification/nutrification - Midpoint	kg SO_2 eq. to air	9.904E-04	9.895E-04	9.396E-04	9.228E-04	8.532E-04	8.558E-04	7.772E-04	8.523E-04	8.549E-04	7.763E-04	8.023E-04	8.049E-04	7.264E-04	8.276E-04	8.302E-04	7.516E-04
	ILCD Acidification midpoint (v1.09)	Mole of H^+ eq.	2.294E-04	2.291E-04	2.215E-04	2.178E-04	1.928E-04	1.921E-04	1.739E-04	1.925E-04	1.919E-04	1.737E-04	1.849E-04	1.843E-04	1.661E-04	1.888E-04	1.882E-04	1.700E-04
	ReCiPe 2016 Midpoint (E) - Terrestrial Acidification	kg SO_2 eq.	1.414E-04	1.412E-04	1.372E-04	1.349E-04	1.043E-04	1.174E-04	1.061E-04	1.178E-04	1.172E-04	1.060E-04	1.138E-04	1.133E-04	1.020E-04	1.159E-04	1.153E-04	1.041E-04
	TRACI 2.1, Acidification	kg SO_2 eq.	2.061E-04	2.059E-04	1.989E-04	1.955E-04	1.547E-04	1.757E-04	1.582E-04	1.740E-04	1.755E-04	1.580E-04	1.670E-04	1.684E-04	1.510E-04	1.706E-04	1.720E-04	1.546E-04
Terrestrial ecotoxicity	CML2001 - Jan. 2016, Terrestric Ecotoxicity Potential (TETP inf.)	kg DCB eq.	1.713E-04	1.712E-04	1.660E-04	1.558E-04	1.372E-04	1.340E-04	1.273E-04	1.371E-04	1.339E-04	1.272E-04	1.319E-04	1.287E-04	1.220E-04	1.350E-04	1.317E-04	1.251E-04
	ReCiPe 2016 Midpoint (H) - Terrestrial ecotoxicity	kg 1,4-DB eq.	2.783E-02	2.782E-02	2.749E-02	2.682E-02	2.000E-02	2.336E-02	2.000E-02	2.083E-02	2.334E-02	1.999E-02	2.051E-02	2.302E-02	1.966E-02	2.068E-02	2.319E-02	1.983E-02

Life-cycle assessment method / Concrete mixture option	4-0-0 (17)2F-0	5-0-0 (19)2S-0	6-0-0 (18)2B-0	1-1-1 (29)1F-1Cs	2-1-1 (31)1S-1Cs	3-1-1 (30)1B-1Cs	1-1-2 (32)1F-1S	2-1-2 (34)1S-1S	3-1-2 (33)1B-1S	4-1-0 (20)2F-0.5W	5-1-0 (22)2S-0.5W	6-1-0 (21)2B-0.5W	4-0-1 (23)2F-0.5Cs	5-0-1 (25)2S-0.5Cs	6-0-1 (24)2B-0.5Cs	4-0-2 (26)2F-0.5S	5-0-2 (28)2S-0.5S	6-0-2 (27)2B-0.5S	4-1-1 (35)2F-1Cs	5-1-1 (37)2S-1Cs	6-1-1 (36)2B-1Cs	4-1-2 (38)2F-1S
Points to achieve	2	2	2	2	2	2	2	2	2	2.5	2.5	2.5	2.5	2.5	2.5	2.5	2.5	2.5	3	3	3	3
CML2001 - Jan. 2016, Global Warming Potential (GWP 100 years)	0.0783	0.0941	0.0786	0.0974	0.1003	0.0887	0.0980	0.1009	0.0893	0.0900	0.0939	0.0939	0.0890	0.0928	0.0773	0.0896	0.0935	0.0780	0.0888	0.0926	0.0771	0.0894
IO2+ v2.1 - Global warming 500yr - Midpoint	0.0776	0.0926	0.0783	0.0965	0.0992	0.0882	0.0971	0.0998	0.0888	0.0889	0.0925	0.0925	0.0877	0.0914	0.0771	0.0883	0.0920	0.0777	0.0877	0.0913	0.0767	0.0883
ILCD - Climate change midpoint, incl biogenic carbon (v1.09)	0.0779	0.0934	0.0782	0.0972	0.0999	0.0885	0.0978	0.1005	0.0891	0.0897	0.0933	0.0933	0.0886	0.0922	0.0770	0.0892	0.0928	0.0776	0.0885	0.0921	0.0769	0.0891
ReCiPe 2016 Midpoint (H) - Climate change, incl biogenic carbon	0.0770	0.0919	0.0774	0.0958	0.0985	0.0876	0.0964	0.0991	0.0882	0.0882	0.0919	0.0919	0.0871	0.0907	0.0762	0.0877	0.0913	0.0768	0.0870	0.0907	0.0761	0.0876
TRACI 2.1, Global Warming Air, incl. biogenic carbon	0.0778	0.0933	0.0782	0.0971	0.0998	0.0884	0.0977	0.1004	0.0891	0.0897	0.0933	0.0933	0.0885	0.0921	0.0770	0.0891	0.0927	0.0776	0.0885	0.0920	0.0769	0.0891
CML2001 - Jan. 2016, Freshwater Aquatic Ecotoxicity Pot. (FAETP inf.)	6.821E-05	8.641E-05	6.885E-05	8.159E-05	8.630E-05	7.316E-05	8.342E-05	8.817E-05	7.498E-05	7.957E-05	8.589E-05	8.589E-05	7.666E-05	8.297E-05	6.542E-05	7.849E-05	8.487E-05	6.724E-05	7.614E-05	8.245E-05	6.489E-05	7.796E-05
IO2+ v2.1 - Aquatic ecotoxicity - Midpoint	1.984E-01	3.260E-01	2.002E-01	2.568E-01	2.923E-01	1.980E-01	2.667E-01	3.023E-01	2.079E-01	2.762E-01	3.236E-01	3.236E-01	2.620E-01	3.094E-01	1.836E-01	2.720E-01	3.195E-01	1.935E-01	2.596E-01	3.070E-01	1.812E-01	2.696E-01
ILCD Ecotoxicity freshwater midpoint (v1.09)	1.589E-03	1.719E-03	1.616E-03	1.930E-03	1.695E-03	1.619E-03	1.999E-03	1.767E-03	1.689E-03	2.008E-03	1.696E-03	1.696E-03	1.905E-03	1.593E-03	1.491E-03	1.974E-03	1.666E-03	1.560E-03	1.882E-03	1.571E-03	1.468E-03	1.951E-03
ReCiPe 2016 Midpoint (E) - Freshwater ecotoxicity	2.849E-06	3.741E-06	2.881E-06	3.508E-06	3.638E-06	2.993E-06	3.593E-06	3.724E-06	3.078E-06	3.525E-06	3.699E-06	3.699E-06	3.406E-06	3.579E-06	2.719E-06	3.491E-06	3.668E-06	2.804E-06	3.364E-06	3.537E-06	2.677E-06	3.449E-06
TRACI 2.1, Ecotoxicity	8.801E-04	8.720E-04	8.889E-04	1.147E-03	8.908E-04	9.036E-04	1.186E-03	9.304E-04	9.427E-04	1.202E-03	8.601E-04	8.601E-04	1.150E-03	8.083E-04	8.251E-04	1.189E-03	8.483E-04	8.642E-04	1.138E-03	7.964E-04	8.132E-04	1.177E-03
CML2001 - Jan. 2016, Eutrophication Potential (EP)	1.751E-05	2.170E-05	1.785E-05	2.018E-05	2.091E-05	1.802E-05	2.080E-05	2.153E-05	1.864E-05	2.051E-05	2.149E-05	2.149E-05	1.949E-05	2.046E-05	1.661E-05	2.011E-05	2.109E-05	1.723E-05	1.928E-05	2.025E-05	1.640E-05	1.990E-05
IO2+ v2.1 - Aquatic eutrophication - Midpoint	1.767E-07	3.439E-07	1.769E-07	1.716E-07	2.941E-07	1.688E-07	1.792E-07	3.017E-07	1.765E-07	1.712E-07	3.346E-07	3.346E-07	1.671E-07	3.304E-07	1.634E-07	1.747E-07	3.380E-07	1.710E-07	1.578E-07	3.211E-07	1.541E-07	1.654E-07
ILCD Eutrophication freshwater midpoint (v1.09)	5.804E-08	1.131E-07	5.810E-08	5.640E-08	9.673E-08	5.550E-08	5.890E-08	9.923E-08	5.800E-08	5.627E-08	1.100E-07	1.100E-07	5.489E-08	1.087E-07	5.368E-08	5.739E-08	1.112E-07	5.619E-08	5.185E-08	1.056E-07	5.064E-08	5.435E-08
ReCiPe 2016 Midpoint (H) - Freshwater Eutrophication	5.800E-08	1.130E-07	5.805E-08	5.634E-08	9.664E-08	5.545E-08	5.884E-08	9.914E-08	5.795E-08	5.621E-08	1.099E-07	1.099E-07	5.483E-08	1.086E-07	5.364E-08	5.733E-08	1.111E-07	5.614E-08	5.179E-08	1.055E-07	5.060E-08	5.429E-08
TRACI 2.1, Eutrophication	9.291E-06	1.270E-05	9.408E-06	1.111E-05	1.205E-05	9.581E-06	1.137E-05	1.231E-05	9.845E-06	1.140E-05	1.265E-05	1.265E-05	1.094E-05	1.219E-05	8.899E-06	1.120E-05	1.246E-05	9.164E-06	1.089E-05	1.214E-05	8.850E-06	1.115E-05
CML2001 - Jan. 2016, Human Toxicity Potential (HTP inf.)	1.621E-02	1.694E-02	1.622E-02	1.912E-02	1.937E-02	1.883E-02	1.915E-02	1.939E-02	1.885E-02	1.661E-02	1.694E-02	1.694E-02	1.657E-02	1.690E-02	1.617E-02	1.659E-02	1.692E-02	1.619E-02	1.656E-02	1.689E-02	1.617E-02	1.659E-02
ILCD Human toxicity midpoint, cancer effects (v1.09)	7.425E-11	8.525E-11	7.609E-11	8.111E-11	8.036E-11	7.355E-11	8.466E-11	8.402E-11	7.710E-11	8.494E-11	8.403E-11	8.403E-11	7.900E-11	7.809E-11	6.893E-11	8.255E-11	8.183E-11	7.248E-11	7.778E-11	7.687E-11	6.771E-11	8.133E-11
ReCiPe 2016 Midpoint (E) - Human toxicity, cancer	2.395E-04	2.917E-04	2.441E-04	2.512E-04	2.600E-04	2.245E-04	2.628E-04	2.719E-04	2.361E-04	2.687E-04	2.807E-04	2.807E-04	2.562E-04	2.681E-04	2.206E-04	2.678E-04	2.802E-04	2.321E-04	2.452E-04	2.571E-04	2.095E-04	2.568E-04
TRACI 2.1, Human toxicity, cancer	2.784E-11	3.174E-11	2.788E-11	3.158E-11	3.263E-11	2.974E-11	3.230E-11	3.335E-11	3.045E-11	3.013E-11	3.154E-11	3.154E-11	2.895E-11	3.036E-11	2.649E-11	2.967E-11	3.108E-11	2.721E-11	2.875E-11	3.015E-11	2.629E-11	2.946E-11
CML2001 - Jan. 2016, Photochem. Ozone Creation Potential (POCP)	6.005E-06	7.698E-06	5.465E-06	8.844E-06	1.017E-05	8.494E-06	8.289E-06	9.615E-06	7.940E-06	5.910E-06	7.677E-06	7.677E-06	7.053E-06	8.820E-06	6.587E-06	6.499E-06	8.269E-06	5.426E-06	7.033E-06	8.800E-06	6.567E-06	6.478E-06
IO2+ v2.1 - Photochemical oxidation - Midpoint	4.717E-06	5.698E-06	4.751E-06	5.561E-06	5.716E-06	5.007E-06	5.691E-06	5.848E-06	5.137E-06	5.473E-06	5.681E-06	5.681E-06	5.227E-06	5.435E-06	4.487E-06	5.357E-06	5.567E-06	4.617E-06	5.210E-06	5.418E-06	4.470E-06	5.340E-06
ILCD - Photochemical ozone formation midpoint, human health (v1.09)	1.294E-04	1.508E-04	1.313E-04	1.507E-04	1.527E-04	1.381E-04	1.546E-04	1.565E-04	1.420E-04	1.480E-04	1.506E-04	1.506E-04	1.404E-04	1.430E-04	1.235E-04	1.442E-04	1.468E-04	1.274E-04	1.402E-04	1.428E-04	1.234E-04	1.441E-04
ReCiPe 2016 Midpoint (H) - Photochemical Ozone Formation, Ecosystems	1.714E-02	4.263E-02	1.717E-02	2.889E-02	3.771E-02	1.861E-02	2.916E-02	3.798E-02	1.888E-02	3.077E-02	4.253E-02	4.253E-02	3.032E-02	4.209E-02	1.662E-02	3.059E-02	4.236E-02	1.689E-02	3.022E-02	4.199E-02	1.652E-02	3.050E-02
CML2001 - Jan. 2016 Ozone Layer Depletion Potential (ODP, steady state)	3.568E-13	3.747E-13	3.568E-13	3.577E-13	3.710E-13	3.575E-13	3.580E-13	3.713E-13	3.579E-13	3.569E-13	3.746E-13	3.746E-13	3.563E-13	3.741E-13	3.561E-13	3.567E-13	3.744E-13	3.565E-13	3.563E-13	3.740E-13	3.561E-13	3.566E-13
IO2+ v2.1 - Ozone layer depletion - Midpoint	3.568E-13	3.747E-13	3.568E-13	3.577E-13	3.710E-13	3.575E-13	3.580E-13	3.713E-13	3.579E-13	3.569E-13	3.746E-13	3.567E-13	3.563E-13	3.741E-13	3.561E-13	3.567E-13	3.744E-13	3.565E-13	3.563E-13	3.740E-13	3.561E-13	3.566E-13
ILCD - Ozone depletion midpoint (v1.09)	3.570E-13	3.752E-13	3.570E-13	3.579E-13	3.714E-13	3.577E-13	3.582E-13	3.717E-13	3.581E-13	3.571E-13	3.751E-13	3.751E-13	3.565E-13	3.745E-13	3.563E-13	3.568E-13	3.748E-13	3.566E-13	3.564E-13	3.744E-13	3.562E-13	3.567E-13
ReCiPe 2016 Midpoint (H) - Stratospheric Ozone Depletion	1.288E-08	2.323E-08	1.301E-08	1.993E-08	2.130E-08	1.364E-08	2.018E-08	2.156E-08	1.389E-08	2.128E-08	2.312E-08	2.312E-08	2.092E-08	2.275E-08	1.253E-08	2.117E-08	2.301E-08	1.278E-08	2.080E-08	2.264E-08	1.242E-08	2.106E-08
TRACI 2.1, Ozone Depletion Air	3.787E-13	3.967E-13	3.787E-13	3.796E-13	3.929E-13	3.795E-13	3.800E-13	3.933E-13	3.798E-13	3.788E-13	3.966E-13	3.966E-13	3.783E-13	3.960E-13	3.781E-13	3.786E-13	3.963E-13	3.784E-13	3.782E-13	3.959E-13	3.780E-13	3.785E-13
CML2001 - Jan. 2016, Acidification Potential (AP)	1.227E-04	1.464E-04	1.241E-04	1.478E-04	1.495E-04	1.328E-04	1.505E-04	1.523E-04	1.356E-04	1.432E-04	1.455E-04	1.455E-04	1.385E-04	1.408E-04	1.186E-04	1.413E-04	1.436E-04	1.213E-04	1.376E-04	1.400E-04	1.177E-04	1.404E-04
IO2+ v2.1 - Terrestrial acidification/nutrification - Midpoint	6.919E-04	8.109E-04	7.061E-04	8.014E-04	8.040E-04	7.254E-04	8.267E-04	8.293E-04	7.507E-04	8.065E-04	8.100E-04	8.100E-04	7.566E-04	7.601E-04	6.553E-04	7.818E-04	7.854E-04	6.805E-04	7.557E-04	7.592E-04	6.544E-04	7.809E-04
ILCD Acidification midpoint (v1.09)	1.522E-04	1.797E-04	1.555E-04	1.847E-04	1.841E-04	1.659E-04	1.886E-04	1.880E-04	1.698E-04	1.803E-04	1.795E-04	1.795E-04	1.727E-04	1.719E-04	1.476E-04	1.766E-04	1.758E-04	1.515E-04	1.725E-04	1.717E-04	1.474E-04	1.764E-04
ReCiPe 2016 Midpoint (E) - Terrestrial Acidification	9.337E-05	1.094E-04	9.440E-05	1.137E-04	1.131E-04	1.019E-04	1.157E-04	1.152E-04	1.039E-04	1.100E-04	1.092E-04	1.092E-04	1.070E-04	1.090E-04	9.027E-05	1.081E-04	1.073E-04	9.233E-05	1.059E-04	1.051E-04	9.012E-05	1.079E-04
TRACI 2.1, Acidification	1.403E-04	1.655E-04	1.422E-04	1.668E-04	1.682E-04	1.508E-04	1.704E-04	1.718E-04	1.543E-04	1.634E-04	1.653E-04	1.653E-04	1.564E-04	1.583E-04	1.350E-04	1.600E-04	1.619E-04	1.386E-04	1.561E-04	1.581E-04	1.348E-04	1.597E-04
CML2001 - Jan. 2016, Terrestric Ecotoxicity Potential (TETP inf.)	1.126E-04	1.215E-04	1.126E-04	1.318E-04	1.286E-04	1.219E-04	1.349E-04	1.317E-04	1.250E-04	1.258E-04	1.214E-04	1.214E-04	1.205E-04	1.162E-04	1.073E-04	1.236E-04	1.193E-04	1.104E-04	1.205E-04	1.161E-04	1.073E-04	1.235E-04
ReCiPe 2016 Midpoint (H) - Terrestrial ecotoxicity	1.739E-02	2.187E-02	1.740E-02	2.049E-02	2.300E-02	1.965E-02	2.066E-02	2.317E-02	1.982E-02	1.851E-02	2.185E-02	2.185E-02	1.818E-02	2.153E-02	1.705E-02	1.835E-02	2.170E-02	1.722E-02	1.817E-02	2.151E-02	1.704E-02	1.834E-02

Table 4.9 Life-cycle greenhouse gas emissions of 1 kg 25 MPa concrete

Impact category	Life-cycle assessment method	Unit	0-0-0	0-1-0	0-0-1	0-0-2	1-0-0	2-0-0	3-0-0	1-1-0	2-1-0	3-1-0	1-0-1	2-0-1	3-0-1	1-0-2	2-0-2	3-0-2
	Concrete mixture option		(1)0-0	(2)0-0.5W	(3)0-0.5Cs	(4)0-0.5S	(5)1F-0	(7)1S-0	(6)1B-0	(8)1F-0.5W	(10)1S-0.5W	(9)1B-0.5W	(11)1F-0.5Cs	(13)1S-0.5Cs	(12)1B-0.5Cs	(14)1F-0.5S	(16)1S-0.5S	(15)1B-0.5S
	Points to achieve			0.5	0.5	0.5	1	1	1	1.5	1.5	1.5	1.5	1.5	1.5	1.5	1.5	1.5
Global Warming Potential (GWP 100 years)	CML2001 - Jan. 2016, Global Warming Potential (GWP 100 years)	kg CO2 eq.	0.1252	0.1246	0.1236	0.1228	0.0989	0.1018	0.0901	0.0986	0.1015	0.0899	0.0976	0.1005	0.0889	0.0982	0.1011	0.0895
	I02+ v2.1 - Global warming 500yr - Midpoint	kg CO2 eq.	0.1248	0.1241	0.1229	0.1225	0.0977	0.1005	0.0895	0.0977	0.1004	0.0894	0.0965	0.0993	0.0886	0.0971	0.0999	0.0892
	ILCD - Climate change midpoint, incl biogenic carbon (v1.09)	kg CO2 eq.	0.1251	0.1244	0.1232	0.1227	0.0984	0.1011	0.0898	0.0984	0.1011	0.0897	0.0972	0.0999	0.0885	0.0978	0.1005	0.0891
	ReCiPe 2016 Midpoint (H) - Climate change, incl biogenic carbon	kg CO2 eq.	0.1239	0.1232	0.1221	0.1216	0.0882	0.0997	0.0888	0.0970	0.0997	0.0888	0.0958	0.0986	0.0876	0.0964	0.0991	0.0882
	TRACI 2.1, Global Warming Air, incl. biogenic carbon	kg CO2 eq.	0.1251	0.1243	0.1232	0.1227	0.0891	0.1011	0.0897	0.0983	0.1010	0.0897	0.0972	0.0999	0.0885	0.0978	0.1005	0.0891
Aquatic Ecotoxicity	CML2001 - Jan. 2016, Freshwater Aquatic Ecotoxicity Pot. (FAETP inf.)	kg DCB eq.	1.024E-04	1.014E-04	9.847E-05	9.463E-05	8.555E-05	9.027E-05	7.712E-05	8.503E-05	8.974E-05	7.659E-05	8.211E-05	8.683E-05	7.368E-05	8.394E-05	8.869E-05	7.551E-05
	I02+ v2.1 - Aquatic ecotoxicity - Midpoint	kg TEG eq. to water	2.683E-01	2.650E-01	2.508E-01	2.260E-01	2.758E-01	3.113E-01	2.170E-01	2.734E-01	3.089E-01	2.146E-01	2.592E-01	2.947E-01	2.004E-01	2.691E-01	3.047E-01	2.103E-01
	ILCD Ecotoxicity freshwater midpoint (v1.09)	CTUe	2.231E-03	2.199E-03	2.096E-03	1.976E-03	2.078E-03	1.844E-03	1.768E-03	2.055E-03	1.821E-03	1.745E-03	1.952E-03	1.718E-03	1.642E-03	2.022E-03	1.789E-03	1.711E-03
	ReCiPe 2016 Midpoint (E) - Freshwater ecotoxicity	kg 1,4 DB eq.	4.165E-06	4.104E-06	3.985E-06	3.775E-06	3.174E-06	3.842E-06	3.197E-06	3.670E-06	3.799E-06	3.155E-06	3.551E-06	3.680E-06	3.036E-06	3.636E-06	3.767E-06	3.120E-06
	TRACI 2.1, Ecotoxicity	CTUe	1.256E-03	1.239E-03	1.187E-03	1.098E-03	9.727E-04	9.664E-04	9.793E-04	1.211E-03	9.545E-04	9.674E-04	1.158E-03	9.026E-04	9.155E-04	1.198E-03	9.423E-04	9.546E-04
Eutrophication Potential (EP),	CML2001 - Jan. 2016, Eutrophication Potential (EP)	kg PO4 eq.	2.408E-05	2.411E-05	2.309E-05	2.226E-05	2.163E-05	2.236E-05	1.947E-05	2.141E-05	2.215E-05	1.926E-05	2.039E-05	2.112E-05	1.823E-05	2.101E-05	2.174E-05	1.885E-05
	I02+ v2.1 - Aquatic eutrophication - Midpoint	kg PO4 eq.	2.366E-07	2.265E-07	2.223E-07	1.789E-07	1.944E-07	3.168E-07	1.916E-07	1.851E-07	3.075E-07	1.828E-07	1.809E-07	3.034E-07	1.781E-07	1.885E-07	3.110E-07	1.858E-07
	ILCD Eutrophication freshwater midpoint (v1.09)	kg PO4 eq.	7.587E-08	7.447E-08	7.309E-08	5.886E-08	6.386E-08	1.042E-07	6.295E-08	6.082E-08	1.011E-07	5.992E-08	5.944E-08	9.977E-08	5.853E-08	6.194E-08	1.023E-07	6.104E-08
	ReCiPe 2016 Midpoint (H) - Freshwater Eutrophication	kg PO4 eq.	7.771E-08	7.440E-08	7.302E-08	5.880E-08	6.286E-08	1.041E-07	6.290E-08	6.076E-08	1.011E-07	5.986E-08	5.938E-08	9.968E-08	5.848E-08	6.188E-08	1.022E-07	6.099E-08
	TRACI 2.1, Eutrophication	kg N eq.	1.238E-05	1.228E-05	1.182E-05	1.144E-05	9.915E-06	1.261E-05	1.014E-05	1.162E-05	1.256E-05	1.009E-05	1.116E-05	1.210E-05	9.630E-06	1.142E-05	1.236E-05	9.895E-06
Human toxicity	CML2001 - Jan. 2016, Human Toxicity Potential (HTP inf.)	kg DCB eq.	2.702E-02	2.685E-02	2.681E-02	2.674E-02	1.917E-02	1.942E-02	1.888E-02	1.917E-02	1.942E-02	1.887E-02	1.913E-02	1.938E-02	1.883E-02	1.915E-02	1.940E-02	1.885E-02
	ILCD Human toxicity midpoint, cancer effects (v1.09)	CTUh	9.980E-11	9.824E-11	9.230E-11	8.692E-11	8.949E-11	8.874E-11	8.193E-11	8.827E-11	8.752E-11	8.071E-11	8.233E-11	8.158E-11	7.477E-11	8.588E-11	8.524E-11	7.832E-11
	ReCiPe 2016 Midpoint (E) - Human toxicity, cancer	kg 1,4-DB eq.	3.048E-04	2.929E-04	2.804E-04	2.394E-04	2.556E-04	2.946E-04	2.591E-04	2.748E-04	2.836E-04	2.480E-04	2.623E-04	2.711E-04	2.355E-04	2.738E-04	2.829E-04	2.471E-04
	TRACI 2.1, Human toxicity, cancer	CTUh	4.188E-11	4.148E-11	4.030E-11	3.739E-11	3.130E-11	3.423E-11	3.133E-11	3.297E-11	3.402E-11	3.113E-11	3.179E-11	3.284E-11	2.994E-11	3.250E-11	3.356E-11	3.066E-11
Photochem. Ozone Creation Potential (POCP)	CML2001 - Jan. 2016, Photochem. Ozone Creation Potential (POCP)	kg C2H4 eq. to air	1.327E-05	1.316E-05	1.430E-05	1.322E-05	7.742E-06	9.067E-06	7.393E-06	7.722E-06	9.046E-06	7.372E-06	8.865E-06	1.019E-05	8.515E-06	8.310E-06	9.636E-06	7.961E-06
	I02+ v2.1 - Photochemical oxidation - Midpoint	kg C2H4 eq. to air	6.927E-06	6.879E-06	6.632E-06	6.274E-06	5.842E-06	5.997E-06	5.287E-06	5.825E-06	5.980E-06	5.270E-06	5.578E-06	5.733E-06	5.023E-06	5.708E-06	5.865E-06	5.154E-06
	ILCD - Photochemical ozone formation midpoint, human health (v1.09)	kg NMVOC eq.	1.913E-04	1.902E-04	1.825E-04	1.782E-04	1.425E-04	1.607E-04	1.461E-04	1.585E-04	1.605E-04	1.459E-04	1.509E-04	1.529E-04	1.383E-04	1.548E-04	1.567E-04	1.421E-04
	ReCiPe 2016 Midpoint (H) - Photochemical Ozone Formation, Ecosystems	kg NOx eq.	2.565E-02	2.543E-02	2.498E-02	2.383E-02	1.923E-02	3.835E-02	1.925E-02	2.944E-02	3.826E-02	1.916E-02	2.899E-02	3.781E-02	1.871E-02	2.926E-02	3.808E-02	1.898E-02
Ozone Layer Depletion Potential	CML2001 - Jan. 2016 Ozone Layer Depletion Potential (ODP, steady state)	kg R-11 eq.	3.628E-13	3.627E-13	3.621E-13	3.627E-13	3.584E-13	3.717E-13	3.583E-13	3.583E-13	3.716E-13	3.582E-13	3.578E-13	3.711E-13	3.576E-13	3.581E-13	3.714E-13	3.580E-13
	I02+ v2.1 - Ozone layer depletion - Midpoint	kg R-11 eq.	3.628E-13	3.627E-13	3.621E-13	3.627E-13	3.584E-13	3.717E-13	3.583E-13	3.583E-13	3.716E-13	3.582E-13	3.578E-13	3.711E-13	3.576E-13	3.581E-13	3.714E-13	3.580E-13
	ILCD - Ozone depletion midpoint (v1.09)	kg CFC-11 eq.	3.631E-13	3.629E-13	3.624E-13	3.629E-13	3.586E-13	3.721E-13	3.585E-13	3.585E-13	3.720E-13	3.584E-13	3.580E-13	3.715E-13	3.578E-13	3.583E-13	3.718E-13	3.581E-13
	ReCiPe 2016 Midpoint (H) - Stratospheric Ozone Depletion	kg CFC-11 eq.	1.784E-08	1.778E-08	1.728E-08	1.695E-08	1.400E-08	2.189E-08	1.423E-08	2.041E-08	2.178E-08	1.412E-08	2.004E-08	2.141E-08	1.375E-08	2.030E-08	2.167E-08	1.400E-08
	TRACI 2.1, Ozone Depletion Air	kg CFC-11 eq.	3.848E-13	3.846E-13	3.840E-13	3.848E-13	3.802E-13	3.937E-13	3.802E-13	3.297E-11	3.936E-13	3.801E-13	3.797E-13	3.930E-13	3.796E-13	3.800E-13	3.933E-13	3.799E-13
Terrestrial Acidification	CML2001 - Jan. 2016, Acidification Potential (AP)	kg SO2 eq.	1.855E-04	1.837E-04	1.790E-04	1.747E-04	1.542E-04	1.559E-04	1.392E-04	1.533E-04	1.551E-04	1.384E-04	1.486E-04	1.504E-04	1.337E-04	1.514E-04	1.531E-04	1.364E-04
	I02+ v2.1 - Terrestrial acidification/nutrification - Midpoint	kg SO2 eq. to air	9.950E-04	9.895E-04	9.396E-04	9.230E-04	8.532E-04	8.558E-04	7.772E-04	8.523E-04	8.549E-04	7.763E-04	8.023E-04	8.049E-04	7.264E-04	8.276E-04	8.302E-04	7.516E-04
	ILCD Acidification midpoint (v1.09)	Mole of H+ eq.	2.305E-04	2.291E-04	2.215E-04	2.178E-04	1.928E-04	1.921E-04	1.739E-04	1.925E-04	1.919E-04	1.737E-04	1.849E-04	1.843E-04	1.661E-04	1.888E-04	1.882E-04	1.700E-04
	ReCiPe 2016 Midpoint (E) - Terrestrial Acidification	kg SO2 eq.	1.421E-04	1.412E-04	1.372E-04	1.350E-04	1.043E-04	1.174E-04	1.061E-04	1.178E-04	1.172E-04	1.060E-04	1.138E-04	1.133E-04	1.020E-04	1.159E-04	1.153E-04	1.041E-04
	TRACI 2.1, Acidification	kg SO2 eq.	2.071E-04	2.059E-04	1.989E-04	1.956E-04	1.547E-04	1.757E-04	1.582E-04	1.740E-04	1.755E-04	1.580E-04	1.670E-04	1.684E-04	1.510E-04	1.706E-04	1.720E-04	1.546E-04
Terrestrial ecotoxicity	CML2001 - Jan. 2016, Terrestric Ecotoxicity Potential (TETP inf.)	kg DCB eq.	1.721E-04	1.712E-04	1.660E-04	1.558E-04	1.372E-04	1.340E-04	1.273E-04	1.371E-04	1.339E-04	1.272E-04	1.319E-04	1.287E-04	1.220E-04	1.350E-04	1.317E-04	1.251E-04
	ReCiPe 2016 Midpoint (H) - Terrestrial ecotoxicity	kg 1,4-DB eq.	2.799E-02	2.782E-02	2.749E-02	2.683E-02	2.000E-02	2.336E-02	2.000E-02	2.083E-02	2.334E-02	1.999E-02	2.051E-02	2.302E-02	1.966E-02	2.068E-02	2.319E-02	1.983E-02

4-0-0	5-0-0	6-0-0	1-1-1	2-1-1	3-1-1	1-1-2	2-1-2	3-1-2	4-1-0	5-1-0	6-1-0	4-0-1	5-0-1	6-0-1	4-0-2	5-0-2	6-0-2	4-1-1	5-1-1	6-1-1	4-1-2
(17)2F-0	(19)2S-0	(18)2B-0	(29)1F-1Cs	(31)1S-1Cs	(30)1B-1Cs	(32)1F-1S	(34)1S-1S	(33)1B-1S	(20)2F-0.5W	(22)2S-0.5W	(21)2B-0.5W	(23)2F-0.5Cs	(25)2S-0.5Cs	(24)2B-0.5Cs	(26)2F-0.5S	(28)2S-0.5S	(27)2B-0.5S	(35)2F-1Cs	(37)2S-1Cs	(36)2B-1Cs	(38)2F-1S
2	2	2	2	2	2	2	2	2	2.5	2.5	2.5	2.5	2.5	2.5	2.5	2.5	2.5	3	3	3	3
0.0783	0.0941	0.0786	0.0974	0.1003	0.0887	0.0980	0.1009	0.0893	0.0900	0.0939	0.0939	0.0890	0.0928	0.0773	0.0896	0.0935	0.0780	0.0888	0.0926	0.0771	0.0894
0.0776	0.0926	0.0783	0.0965	0.0992	0.0882	0.0971	0.0998	0.0888	0.0889	0.0925	0.0925	0.0877	0.0914	0.0771	0.0883	0.0920	0.0777	0.0877	0.0913	0.0767	0.0883
0.0779	0.0934	0.0782	0.0972	0.0999	0.0885	0.0978	0.1005	0.0891	0.0897	0.0933	0.0933	0.0886	0.0922	0.0770	0.0892	0.0928	0.0776	0.0885	0.0921	0.0769	0.0891
0.0770	0.0919	0.0774	0.0958	0.0985	0.0876	0.0964	0.0991	0.0882	0.0882	0.0919	0.0919	0.0871	0.0907	0.0762	0.0877	0.0913	0.0768	0.0870	0.0907	0.0761	0.0876
0.0778	0.0933	0.0782	0.0971	0.0998	0.0884	0.0977	0.1004	0.0891	0.0897	0.0933	0.0933	0.0885	0.0921	0.0770	0.0891	0.0927	0.0776	0.0885	0.0920	0.0769	0.0891
6.821E-05	8.641E-05	6.885E-05	8.159E-05	8.630E-05	7.316E-05	8.342E-05	8.817E-05	7.498E-05	7.957E-05	8.589E-05	8.589E-05	7.666E-05	8.297E-05	6.542E-05	7.849E-05	8.487E-05	6.724E-05	7.614E-05	8.245E-05	6.489E-05	7.796E-05
1.984E-01	3.260E-01	2.002E-01	2.568E-01	2.923E-01	1.980E-01	2.667E-01	3.023E-01	2.079E-01	2.762E-01	3.236E-01	3.236E-01	2.620E-01	3.094E-01	1.836E-01	2.720E-01	3.195E-01	1.935E-01	2.596E-01	3.070E-01	1.812E-01	2.696E-01
1.589E-03	1.719E-03	1.616E-03	1.930E-03	1.695E-03	1.619E-03	1.999E-03	1.767E-03	1.689E-03	2.008E-03	1.696E-03	1.696E-03	1.905E-03	1.593E-03	1.491E-03	1.974E-03	1.666E-03	1.560E-03	1.882E-03	1.571E-03	1.468E-03	1.951E-03
2.849E-06	3.741E-06	2.881E-06	3.508E-06	3.638E-06	2.993E-06	3.593E-06	3.724E-06	3.078E-06	3.525E-06	3.699E-06	3.699E-06	3.406E-06	3.579E-06	2.719E-06	3.491E-06	3.668E-06	2.804E-06	3.364E-06	3.537E-06	2.677E-06	3.449E-06
8.801E-04	8.720E-04	8.889E-04	1.147E-03	8.908E-04	9.036E-04	1.186E-03	9.304E-04	9.427E-04	1.202E-03	8.601E-04	8.601E-04	1.150E-03	8.083E-04	8.251E-04	1.189E-03	8.483E-04	8.642E-04	1.138E-03	7.964E-04	8.132E-04	1.177E-03
1.751E-05	2.170E-05	1.785E-05	2.018E-05	2.091E-05	1.802E-05	2.080E-05	2.153E-05	1.864E-05	2.051E-05	2.149E-05	2.149E-05	1.949E-05	2.046E-05	1.661E-05	2.011E-05	2.109E-05	1.723E-05	1.928E-05	2.025E-05	1.640E-05	1.990E-05
1.767E-07	3.439E-07	1.769E-07	1.716E-07	2.941E-07	1.688E-07	1.792E-07	3.017E-07	1.765E-07	1.712E-07	3.346E-07	3.346E-07	1.671E-07	3.304E-07	1.634E-07	1.747E-07	3.380E-07	1.710E-07	1.578E-07	3.211E-07	1.541E-07	1.654E-07
5.804E-08	1.131E-07	5.810E-08	5.640E-08	9.673E-08	5.550E-08	5.890E-08	9.923E-08	5.800E-08	5.627E-08	1.100E-07	1.100E-07	5.489E-08	1.087E-07	5.368E-08	5.739E-08	1.112E-07	5.619E-08	5.185E-08	1.056E-07	5.064E-08	5.435E-08
5.800E-08	1.130E-07	5.805E-08	5.634E-08	9.664E-08	5.545E-08	5.884E-08	9.914E-08	5.795E-08	5.621E-08	1.099E-07	1.099E-07	5.483E-08	1.086E-07	5.364E-08	5.733E-08	1.111E-07	5.614E-08	5.179E-08	1.055E-07	5.060E-08	5.429E-08
9.291E-06	1.270E-05	9.408E-06	1.111E-05	1.205E-05	9.581E-06	1.137E-05	1.231E-05	9.845E-06	1.140E-05	1.265E-05	1.265E-05	1.094E-05	1.219E-05	8.899E-06	1.120E-05	1.246E-05	9.164E-06	1.089E-05	1.214E-05	8.850E-06	1.115E-05
1.621E-02	1.694E-02	1.622E-02	1.912E-02	1.937E-02	1.883E-02	1.915E-02	1.939E-02	1.885E-02	1.661E-02	1.694E-02	1.694E-02	1.657E-02	1.690E-02	1.617E-02	1.659E-02	1.692E-02	1.619E-02	1.656E-02	1.689E-02	1.617E-02	1.659E-02
7.425E-11	8.525E-11	7.609E-11	8.111E-11	8.036E-11	7.355E-11	8.466E-11	8.402E-11	7.710E-11	8.494E-11	8.403E-11	8.403E-11	7.900E-11	7.809E-11	6.893E-11	8.255E-11	8.183E-11	7.248E-11	7.778E-11	7.687E-11	6.771E-11	8.133E-11
2.395E-04	2.917E-04	2.441E-04	2.512E-04	2.600E-04	2.245E-04	2.628E-04	2.719E-04	2.361E-04	2.687E-04	2.807E-04	2.807E-04	2.562E-04	2.681E-04	2.206E-04	2.678E-04	2.802E-04	2.321E-04	2.452E-04	2.571E-04	2.095E-04	2.568E-04
2.784E-11	3.174E-11	2.788E-11	3.158E-11	3.263E-11	2.974E-11	3.230E-11	3.335E-11	3.045E-11	3.013E-11	3.154E-11	3.154E-11	2.895E-11	3.036E-11	2.649E-11	2.967E-11	3.108E-11	2.721E-11	2.875E-11	3.015E-11	2.629E-11	2.946E-11
6.005E-06	7.698E-06	5.465E-06	8.844E-06	1.017E-05	8.494E-06	8.289E-06	9.615E-06	7.940E-06	5.910E-06	7.677E-06	7.677E-06	7.053E-06	8.820E-06	6.587E-06	6.499E-06	8.269E-06	5.426E-06	7.033E-06	8.800E-06	6.567E-06	6.478E-06
4.717E-06	5.698E-06	4.751E-06	5.561E-06	5.716E-06	5.007E-06	5.691E-06	5.848E-06	5.137E-06	5.473E-06	5.681E-06	5.681E-06	5.227E-06	5.435E-06	4.487E-06	5.357E-06	5.567E-06	4.617E-06	5.210E-06	5.418E-06	4.470E-06	5.340E-06
1.294E-04	1.508E-04	1.313E-04	1.507E-04	1.527E-04	1.381E-04	1.546E-04	1.565E-04	1.420E-04	1.480E-04	1.506E-04	1.506E-04	1.404E-04	1.430E-04	1.235E-04	1.442E-04	1.468E-04	1.274E-04	1.402E-04	1.428E-04	1.234E-04	1.441E-04
1.714E-02	4.263E-02	1.717E-02	2.889E-02	3.771E-02	1.861E-02	2.916E-02	3.798E-02	1.888E-02	3.077E-02	4.253E-02	4.253E-02	3.032E-02	4.209E-02	1.662E-02	3.059E-02	4.236E-02	1.689E-02	3.022E-02	4.199E-02	1.652E-02	3.050E-02
3.568E-13	3.747E-13	3.568E-13	3.577E-13	3.710E-13	3.575E-13	3.580E-13	3.713E-13	3.579E-13	3.569E-13	3.746E-13	3.746E-13	3.563E-13	3.741E-13	3.561E-13	3.567E-13	3.744E-13	3.565E-13	3.563E-13	3.740E-13	3.561E-13	3.566E-13
3.568E-13	3.747E-13	3.568E-13	3.577E-13	3.710E-13	3.575E-13	3.580E-13	3.713E-13	3.579E-13	3.569E-13	3.746E-13	3.567E-13	3.563E-13	3.741E-13	3.561E-13	3.567E-13	3.744E-13	3.565E-13	3.563E-13	3.740E-13	3.561E-13	3.566E-13
3.570E-13	3.752E-13	3.570E-13	3.579E-13	3.714E-13	3.577E-13	3.582E-13	3.717E-13	3.581E-13	3.571E-13	3.751E-13	3.751E-13	3.565E-13	3.745E-13	3.563E-13	3.568E-13	3.748E-13	3.566E-13	3.564E-13	3.744E-13	3.562E-13	3.567E-13
1.288E-08	2.323E-08	1.301E-08	1.993E-08	2.130E-08	1.364E-08	2.018E-08	2.156E-08	1.389E-08	2.128E-08	2.312E-08	2.312E-08	2.092E-08	2.275E-08	1.253E-08	2.117E-08	2.301E-08	1.278E-08	2.080E-08	2.264E-08	1.242E-08	2.106E-08
3.787E-13	3.967E-13	3.787E-13	3.796E-13	3.929E-13	3.795E-13	3.800E-13	3.933E-13	3.798E-13	3.788E-13	3.966E-13	3.966E-13	3.783E-13	3.960E-13	3.781E-13	3.786E-13	3.963E-13	3.784E-13	3.782E-13	3.959E-13	3.780E-13	3.785E-13
1.227E-04	1.464E-04	1.241E-04	1.478E-04	1.495E-04	1.328E-04	1.505E-04	1.523E-04	1.356E-04	1.432E-04	1.455E-04	1.455E-04	1.385E-04	1.408E-04	1.186E-04	1.413E-04	1.436E-04	1.213E-04	1.376E-04	1.400E-04	1.177E-04	1.404E-04
6.919E-04	8.109E-04	7.061E-04	8.014E-04	8.040E-04	7.254E-04	8.267E-04	8.293E-04	7.507E-04	8.065E-04	8.100E-04	8.100E-04	7.566E-04	7.601E-04	6.553E-04	7.818E-04	7.854E-04	6.805E-04	7.557E-04	7.592E-04	6.544E-04	7.809E-04
1.522E-04	1.797E-04	1.555E-04	1.847E-04	1.841E-04	1.659E-04	1.886E-04	1.880E-04	1.698E-04	1.803E-04	1.795E-04	1.795E-04	1.727E-04	1.719E-04	1.476E-04	1.766E-04	1.758E-04	1.515E-04	1.725E-04	1.717E-04	1.474E-04	1.764E-04
9.337E-05	1.094E-04	9.440E-05	1.137E-04	1.131E-04	1.019E-04	1.157E-04	1.152E-04	1.039E-04	1.100E-04	1.092E-04	1.092E-04	1.070E-04	1.050E-04	9.027E-05	1.081E-04	1.073E-04	9.233E-05	1.059E-04	1.051E-04	9.012E-05	1.079E-04
1.403E-04	1.655E-04	1.422E-04	1.668E-04	1.682E-04	1.508E-04	1.704E-04	1.718E-04	1.543E-04	1.634E-04	1.653E-04	1.653E-04	1.564E-04	1.583E-04	1.350E-04	1.600E-04	1.619E-04	1.386E-04	1.561E-04	1.581E-04	1.348E-04	1.597E-04
1.126E-04	1.215E-04	1.126E-04	1.318E-04	1.286E-04	1.219E-04	1.349E-04	1.317E-04	1.250E-04	1.258E-04	1.214E-04	1.214E-04	1.205E-04	1.162E-04	1.073E-04	1.236E-04	1.193E-04	1.104E-04	1.205E-04	1.161E-04	1.073E-04	1.235E-04
1.739E-02	2.187E-02	1.740E-02	2.049E-02	2.300E-02	1.965E-02	2.066E-02	2.317E-02	1.982E-02	1.851E-02	2.185E-02	2.185E-02	1.818E-02	2.153E-02	1.705E-02	1.835E-02	2.170E-02	1.722E-02	1.817E-02	2.151E-02	1.704E-02	1.834E-02

Table 4.10 Life-cycle greenhouse gas emissions of 1 kg 32 MPa concrete

Impact category	Life-cycle assessment method	Unit	0-0-0	0-1-0	0-0-1	0-0-2	1-0-0	2-0-0	3-0-0	1-1-0	2-1-0	3-1-0	1-0-1	2-0-1	3-0-1	1-0-2	2-0-2	3-0-2
	Concrete mixture option		(1)0-0	(2)0-0.5W	(3)0-0.5Cs	(4)0-0.5S	(5)1F-0	(7)1S-0	(6)1B-0	(8)1F-0.5W	(10)1S-0.5W	(9)1B-0.5W	(11)1F-0.5Cs	(13)1S-0.5Cs	(12)1B-0.5Cs	(14)1F-0.5S	(16)1S-0.5S	(15)1B-0.5S
	Points to achieve		0	0.5	0.5	0.5	1	1	1	1.5	1.5	1.5	1.5	1.5	1.5	1.5	1.5	1.5
Global Warming Potential (GWP 100 years)	CML2001 - Jan. 2016, Global Warming Potential (GWP 100 years)	kg CO_2 eq.	0.1280	0.1281	0.1270	0.1228	0.1016	0.1046	0.0926	0.1014	0.1043	0.0924	0.1003	0.1033	0.0913	0.1010	0.1040	0.0920
	I02+ v2.1 - Global warming 500yr - Midpoint	kg CO_2 eq.	0.1276	0.1276	0.1264	0.1225	0.1004	0.1032	0.0919	0.1004	0.1032	0.0919	0.0992	0.1020	0.0911	0.0998	0.1026	0.0917
	ILCD - Climate change midpoint, incl biogenic carbon (v1.09)	kg CO_2 eq.	0.1279	0.1279	0.1267	0.1228	0.1012	0.1040	0.0922	0.1011	0.1039	0.0922	0.0999	0.1027	0.0910	0.1006	0.1033	0.0916
	ReCiPe 2016 Midpoint (H) - Climate change, incl biogenic carbon	kg CO_2 eq.	0.1267	0.1267	0.1255	0.1217	0.0906	0.1025	0.0913	0.0997	0.1025	0.0912	0.0985	0.1013	0.0901	0.0991	0.1019	0.0907
	TRACI 2.1, Global Warming Air, incl. biogenic carbon	kg CO_2 eq.	0.1279	0.1278	0.1266	0.1228	0.0915	0.1039	0.0922	0.1011	0.1038	0.0921	0.0999	0.1026	0.0909	0.1005	0.1033	0.0916
Aquatic Ecotoxicity	CML2001 - Jan. 2016, Freshwater Aquatic Ecotoxicity Pot. (FAETP inf.)	kg DCB eq.	1.043E-04	1.038E-04	1.009E-04	9.468E-05	8.750E-05	9.236E-05	7.881E-05	8.698E-05	9.183E-05	7.829E-05	8.406E-05	8.892E-05	7.538E-05	8.591E-05	9.080E-05	7.722E-05
	I02+ v2.1 - Aquatic ecotoxicity - Midpoint	kg TEG eq. to water	2.722E-01	2.698E-01	2.556E-01	2.265E-01	2.809E-01	3.174E-01	2.203E-01	2.785E-01	3.150E-01	2.179E-01	2.643E-01	3.008E-01	2.037E-01	2.743E-01	3.109E-01	2.137E-01
	ILCD Ecotoxicity freshwater midpoint (v1.09)	CTUe	2.266E-03	2.243E-03	2.140E-03	1.977E-03	2.118E-03	1.877E-03	1.799E-03	2.096E-03	1.854E-03	1.776E-03	1.993E-03	1.751E-03	1.673E-03	2.063E-03	1.823E-03	1.743E-03
	ReCiPe 2016 Midpoint (E) - Freshwater ecotoxicity	kg 1,4 DB eq.	4.239E-06	4.197E-06	4.078E-06	3.778E-06	3.238E-06	3.926E-06	3.262E-06	3.750E-06	3.883E-06	3.219E-06	3.631E-06	3.764E-06	3.100E-06	3.717E-06	3.852E-06	3.186E-06
	TRACI 2.1, Ecotoxicity	CTUe	1.277E-03	1.265E-03	1.213E-03	1.099E-03	9.909E-04	9.844E-04	9.977E-04	1.237E-03	9.725E-04	9.858E-04	1.185E-03	9.206E-04	9.339E-04	1.224E-03	9.606E-04	9.733E-04
Eutrophication Potential (EP)	CML2001 - Jan. 2016, Eutrophication Potential (EP)	kg PO_4 eq.	2.451E-05	2.464E-05	2.359E-05	2.230E-05	2.208E-05	2.284E-05	1.986E-05	2.187E-05	2.262E-05	1.965E-05	2.082E-05	2.158E-05	1.860E-05	2.147E-05	2.222E-05	1.925E-05
	I02+ v2.1 - Aquatic eutrophication - Midpoint	kg PO_4 eq.	2.399E-07	2.306E-07	2.265E-07	1.792E-07	1.972E-07	3.234E-07	1.944E-07	1.880E-07	3.141E-07	1.851E-07	1.838E-07	3.099E-07	1.810E-07	1.915E-07	3.176E-07	1.886E-07
	ILCD Eutrophication freshwater midpoint (v1.09)	kg PO_4 eq.	7.698E-08	7.583E-08	7.445E-08	5.895E-08	6.481E-08	1.063E-07	6.388E-08	6.177E-08	1.033E-07	6.084E-08	6.039E-08	1.019E-07	5.948E-08	6.292E-08	1.045E-07	6.199E-08
	ReCiPe 2016 Midpoint (H) - Freshwater Eutrophication	kg PO_4 eq.	7.880E-08	7.576E-08	7.438E-08	5.890E-08	6.378E-08	1.062E-07	6.383E-08	6.171E-08	1.032E-07	6.079E-08	6.033E-08	1.018E-07	5.941E-08	6.286E-08	1.044E-07	6.193E-08
	TRACI 2.1, Eutrophication	kg N eq.	1.256E-05	1.251E-05	1.205E-05	1.150E-05	1.008E-05	1.285E-05	1.031E-05	1.183E-05	1.280E-05	1.026E-05	1.137E-05	1.233E-05	9.792E-06	1.164E-05	1.261E-05	1.007E-05
Human toxicity	CML2001 - Jan. 2016, Human Toxicity Potential (HTP inf.)	kg DCB eq.	2.765E-02	2.765E-02	2.760E-02	2.674E-02	1.974E-02	1.999E-02	1.943E-02	1.973E-02	1.999E-02	1.943E-02	1.969E-02	1.995E-02	1.939E-02	1.971E-02	1.997E-02	1.941E-02
	ILCD Human toxicity midpoint, cancer effects (v1.09)	CTUh	1.012E-10	9.995E-11	9.401E-11	8.698E-11	9.090E-11	9.013E-11	8.312E-11	8.968E-11	8.891E-11	8.190E-11	8.374E-11	8.297E-11	7.596E-11	8.733E-11	8.668E-11	7.955E-11
	ReCiPe 2016 Midpoint (E) - Human toxicity, cancer	kg 1,4-DB eq.	3.082E-04	2.971E-04	2.846E-04	2.398E-04	2.584E-04	2.986E-04	2.620E-04	2.785E-04	2.875E-04	2.509E-04	2.660E-04	2.750E-04	2.384E-04	2.777E-04	2.870E-04	2.501E-04
	TRACI 2.1, Human toxicity, cancer	CTUh	4.268E-11	4.248E-11	4.130E-11	3.741E-11	3.199E-11	3.500E-11	3.202E-11	3.371E-11	3.480E-11	3.182E-11	3.253E-11	3.362E-11	3.064E-11	3.326E-11	3.434E-11	3.136E-11
Photochem. Ozone Creation Potential (POCP)	CML2001 - Jan. 2016, Photochem. Ozone Creation Potential (POCP)	kg C_2H_4 eq. to air	1.365E-05	1.363E-05	1.481E-05	1.321E-05	8.039E-06	9.403E-06	7.679E-06	8.018E-06	9.382E-06	7.658E-06	9.202E-06	1.057E-05	8.842E-06	8.608E-06	9.974E-06	8.248E-06
	I02+ v2.1 - Photochemical oxidation - Midpoint	kg C_2H_4 eq. to air	7.054E-06	7.037E-06	6.789E-06	6.280E-06	5.968E-06	6.128E-06	5.397E-06	5.951E-06	6.111E-06	5.380E-06	5.704E-06	5.864E-06	5.133E-06	5.836E-06	5.998E-06	5.265E-06
	ILCD - Photochemical ozone formation midpoint, human health (v1.09)	kg NMVOC eq.	1.950E-04	1.948E-04	1.870E-04	1.783E-04	1.457E-04	1.645E-04	1.494E-04	1.623E-04	1.643E-04	1.493E-04	1.545E-04	1.565E-04	1.415E-04	1.585E-04	1.605E-04	1.455E-04
	ReCiPe 2016 Midpoint (H) - Photochemical Ozone Formation, Ecosystems	kg NO_x eq.	2.614E-02	2.604E-02	2.559E-02	2.388E-02	1.966E-02	3.935E-02	1.968E-02	3.017E-02	3.925E-02	1.958E-02	2.972E-02	3.880E-02	1.913E-02	2.999E-02	3.908E-02	1.941E-02
Ozone Layer Depletion Potential	CML2001 - Jan. 2016 Ozone Layer Depletion Potential (ODP, steady state)	kg R-11 eq.	3.632E-13	3.631E-13	3.625E-13	3.709E-13	3.587E-13	3.724E-13	3.586E-13	3.586E-13	3.723E-13	3.585E-13	3.581E-13	3.718E-13	3.579E-13	3.584E-13	3.721E-13	3.583E-13
	I02+ v2.1 - Ozone layer depletion - Midpoint	kg R-11 eq.	3.632E-13	3.631E-13	3.625E-13	3.709E-13	3.587E-13	3.724E-13	3.586E-13	3.586E-13	3.723E-13	3.585E-13	3.581E-13	3.718E-13	3.579E-13	3.584E-13	3.721E-13	3.583E-13
	ILCD - Ozone depletion midpoint (v1.09)	kg CFC-11 eq.	3.634E-13	3.634E-13	3.628E-13	3.712E-13	3.589E-13	3.728E-13	3.588E-13	3.588E-13	3.727E-13	3.587E-13	3.583E-13	3.722E-13	3.581E-13	3.586E-13	3.725E-13	3.584E-13
	ReCiPe 2016 Midpoint (H) - Stratospheric Ozone Depletion	kg CFC-11 eq.	1.814E-08	1.816E-08	1.765E-08	1.709E-08	1.427E-08	2.240E-08	1.450E-08	2.087E-08	2.228E-08	1.439E-08	2.049E-08	2.191E-08	1.401E-08	2.076E-08	2.217E-08	1.428E-08
	TRACI 2.1, Ozone Depletion Air	kg CFC-11 eq.	3.851E-13	3.890E-13	3.845E-13	3.935E-13	3.805E-13	3.944E-13	3.805E-13	3.371E-11	3.943E-13	3.804E-13	3.800E-13	3.937E-13	3.799E-13	3.803E-13	3.940E-13	3.802E-13
Terrestrial Acidification	CML2001 - Jan. 2016, Acidification Potential (AP)	kg SO_2 eq.	1.892E-04	1.884E-04	1.836E-04	1.748E-04	1.580E-04	1.598E-04	1.426E-04	1.571E-04	1.589E-04	1.417E-04	1.523E-04	1.541E-04	1.369E-04	1.552E-04	1.570E-04	1.398E-04
	I02+ v2.1 - Terrestrial acidification/nutrification - Midpoint	kg SO_2 eq. to air	1.014E-03	1.013E-03	9.618E-04	9.237E-04	8.726E-04	8.753E-04	7.944E-04	8.717E-04	8.744E-04	7.935E-04	8.208E-04	8.235E-04	7.426E-04	8.471E-04	8.498E-04	7.689E-04
	ILCD Acidification midpoint (v1.09)	Mole of H+ eq.	2.351E-04	2.349E-04	2.272E-04	2.180E-04	1.975E-04	1.968E-04	1.781E-04	1.972E-04	1.966E-04	1.779E-04	1.895E-04	1.889E-04	1.701E-04	1.936E-04	1.929E-04	1.742E-04
	ReCiPe 2016 Midpoint (E) - Terrestrial Acidification	kg SO_2 eq.	1.450E-04	1.448E-04	1.408E-04	1.350E-04	1.068E-04	1.203E-04	1.087E-04	1.208E-04	1.202E-04	1.086E-04	1.167E-04	1.161E-04	1.045E-04	1.188E-04	1.183E-04	1.067E-04
	TRACI 2.1, Acidification	kg SO_2 eq.	2.111E-04	2.109E-04	2.037E-04	1.958E-04	1.581E-04	1.798E-04	1.618E-04	1.781E-04	1.796E-04	1.616E-04	1.710E-04	1.724E-04	1.545E-04	1.747E-04	1.762E-04	1.582E-04
Terrestrial ecotoxicity	CML2001 - Jan. 2016, Terrestric Ecotoxicity Potential (TETP inf.)	kg DCB eq.	1.755E-04	1.755E-04	1.702E-04	1.559E-04	1.404E-04	1.371E-04	1.303E-04	1.404E-04	1.370E-04	1.302E-04	1.351E-04	1.318E-04	1.250E-04	1.383E-04	1.349E-04	1.281E-04
	ReCiPe 2016 Midpoint (H) - Terrestrial ecotoxicity	kg 1,4-DB eq.	2.860E-02	2.859E-02	2.826E-02	2.683E-02	2.054E-02	2.400E-02	2.054E-02	2.140E-02	2.398E-02	2.053E-02	2.107E-02	2.366E-02	2.020E-02	2.125E-02	2.383E-02	2.037E-02

Impact category	Life-cycle assessment method	4-0-0	5-0-0	6-0-0	1-1-1	2-1-1	3-1-1	1-1-2	2-1-2	3-1-2	4-1-0	5-1-0	6-1-0	4-0-1	5-0-1	6-0-1	4-0-2	5-0-2	6-0-2	4-1-1	5-1-1	6-1-1	4-1-2
	Concrete mixture option	(17)2F-0	(19)2S-0	(18)2B-0	(29)1F-1Cs	(31)1S-1Cs	(30)1B-1Cs	(32)1F-1S	(34)1S-1S	(33)1B-1S	(20)2F-0.5W	(22)2S-0.5W	(21)2B-0.5W	(23)2F-0.5Cs	(25)2S-0.5Cs	(24)2B-0.5Cs	(26)2F-0.5S	(28)2S-0.5S	(27)2B-0.5S	(35)2F-1Cs	(37)2S-1Cs	(36)2B-1Cs	(38)2F-1S
	Points to achieve	2	2	2	2	2	2	2	2	2	2.5	2.5	2.5	2.5	2.5	2.5	2.5	2.5	2.5	3	3	3	3
Global Warming Potential (GWP 100 years)	CML2001 - Jan. 2016, Global Warming Potential (GWP 100 years)	0.0804	0.0967	0.0807	0.1001	0.1031	0.0911	0.1007	0.1037	0.0918	0.0925	0.0964	0.0964	0.0914	0.0954	0.0794	0.0921	0.0960	0.0801	0.0912	0.0952	0.0792	0.0918
	I02+ v2.1 - Global warming 500yr - Midpoint	0.0797	0.0951	0.0804	0.0992	0.1020	0.0907	0.0998	0.1026	0.0913	0.0913	0.0951	0.0951	0.0902	0.0939	0.0792	0.0908	0.0945	0.0798	0.0901	0.0939	0.0788	0.0907
	ILCD - Climate change midpoint, incl biogenic carbon (v1.09)	0.0800	0.0960	0.0803	0.0999	0.1026	0.0909	0.1005	0.1033	0.0916	0.0922	0.0959	0.0959	0.0910	0.0947	0.0791	0.0916	0.0954	0.0797	0.0909	0.0947	0.0790	0.0916
	ReCiPe 2016 Midpoint (H) - Climate change, incl biogenic carbon	0.0791	0.0945	0.0795	0.0984	0.1013	0.0900	0.0991	0.1019	0.0906	0.0907	0.0944	0.0944	0.0895	0.0932	0.0783	0.0901	0.0939	0.0789	0.0894	0.0932	0.0782	0.0901
	TRACI 2.1, Global Warming Air, incl. biogenic carbon	0.0800	0.0959	0.0803	0.0998	0.1026	0.0909	0.1005	0.1032	0.0915	0.0922	0.0958	0.0958	0.0910	0.0947	0.0791	0.0916	0.0953	0.0797	0.0909	0.0946	0.0790	0.0915
Aquatic Ecotoxicity	CML2001 - Jan. 2016, Freshwater Aquatic Ecotoxicity Pot. (FAETP inf.)	6.965E-05	8.839E-05	7.030E-05	8.354E-05	8.840E-05	7.485E-05	8.539E-05	9.028E-05	7.670E-05	8.136E-05	8.786E-05	8.786E-05	7.845E-05	8.495E-05	6.687E-05	8.029E-05	8.686E-05	6.871E-05	7.792E-05	8.442E-05	6.634E-05	7.977E-05
	I02+ v2.1 - Aquatic ecotoxicity - Midpoint	2.011E-01	3.326E-01	2.031E-01	2.619E-01	2.984E-01	2.013E-01	2.719E-01	3.085E-01	2.113E-01	2.814E-01	3.302E-01	3.302E-01	2.672E-01	3.160E-01	1.865E-01	2.772E-01	3.262E-01	1.965E-01	2.648E-01	3.136E-01	1.841E-01	2.748E-01
	ILCD Ecotoxicity freshwater midpoint (v1.09)	1.615E-03	1.749E-03	1.643E-03	1.970E-03	1.729E-03	1.650E-03	2.040E-03	1.801E-03	1.720E-03	2.046E-03	1.726E-03	1.726E-03	1.943E-03	1.623E-03	1.517E-03	2.014E-03	1.696E-03	1.587E-03	1.921E-03	1.600E-03	1.494E-03	1.991E-03
	ReCiPe 2016 Midpoint (E) - Freshwater ecotoxicity	2.904E-06	3.822E-06	2.936E-06	3.589E-06	3.722E-06	3.058E-06	3.674E-06	3.809E-06	3.144E-06	3.601E-06	3.780E-06	3.780E-06	3.482E-06	3.660E-06	2.774E-06	3.568E-06	3.750E-06	2.860E-06	3.439E-06	3.618E-06	2.732E-06	3.525E-06
	TRACI 2.1, Ecotoxicity	8.955E-04	8.872E-04	9.046E-04	1.173E-03	9.087E-04	9.220E-04	1.212E-03	9.487E-04	9.614E-04	1.227E-03	8.753E-04	8.753E-04	1.175E-03	8.235E-04	8.408E-04	1.215E-03	8.638E-04	8.802E-04	1.163E-03	8.116E-04	8.289E-04	1.203E-03
Eutrophication Potential (EP)	CML2001 - Jan. 2016, Eutrophication Potential (EP)	1.785E-05	2.216E-05	1.820E-05	2.061E-05	2.136E-05	1.839E-05	2.126E-05	2.201E-05	1.904E-05	2.095E-05	2.195E-05	2.195E-05	1.990E-05	2.090E-05	1.694E-05	2.055E-05	2.155E-05	1.758E-05	1.969E-05	2.069E-05	1.673E-05	2.033E-05
	I02+ v2.1 - Aquatic eutrophication - Midpoint	1.791E-07	3.512E-07	1.793E-07	1.745E-07	3.006E-07	1.717E-07	1.822E-07	3.083E-07	1.794E-07	1.737E-07	3.419E-07	3.419E-07	1.695E-07	3.378E-07	1.658E-07	1.772E-07	3.455E-07	1.735E-07	1.603E-07	3.285E-07	1.565E-07	1.679E-07
	ILCD Eutrophication freshwater midpoint (v1.09)	5.882E-08	1.155E-07	5.888E-08	5.735E-08	9.889E-08	5.642E-08	5.988E-08	1.014E-07	5.895E-08	5.709E-08	1.125E-07	1.125E-07	5.571E-08	1.111E-07	5.447E-08	5.823E-08	1.136E-07	5.699E-08	5.267E-08	1.080E-07	5.143E-08	5.519E-08
	ReCiPe 2016 Midpoint (H) - Freshwater Eutrophication	5.877E-08	1.154E-07	5.884E-08	5.729E-08	9.879E-08	5.637E-08	5.982E-08	1.013E-07	5.890E-08	5.703E-08	1.124E-07	1.124E-07	5.565E-08	1.110E-07	5.442E-08	5.817E-08	1.135E-07	5.694E-08	5.261E-08	1.079E-07	5.138E-08	5.513E-08
	TRACI 2.1, Eutrophication	9.437E-06	1.295E-05	9.557E-06	1.132E-05	1.229E-05	9.743E-06	1.159E-05	1.256E-05	1.002E-05	1.161E-05	1.290E-05	1.290E-05	1.114E-05	1.243E-05	9.041E-06	1.141E-05	1.270E-05	9.314E-06	1.109E-05	1.238E-05	8.991E-06	1.136E-05
Human toxicity	CML2001 - Jan. 2016, Human Toxicity Potential (HTP inf.)	1.669E-02	1.744E-02	1.669E-02	1.969E-02	1.994E-02	1.938E-02	1.971E-02	1.997E-02	1.941E-02	1.710E-02	1.744E-02	1.744E-02	1.705E-02	1.739E-02	1.665E-02	1.708E-02	1.742E-02	1.667E-02	1.705E-02	1.739E-02	1.664E-02	1.707E-02
	ILCD Human toxicity midpoint, cancer effects (v1.09)	7.521E-11	8.653E-11	7.710E-11	8.252E-11	8.175E-11	7.474E-11	8.611E-11	8.546E-11	7.833E-11	8.626E-11	8.531E-11	8.531E-11	8.032E-11	7.937E-11	6.994E-11	8.391E-11	8.317E-11	7.353E-11	7.910E-11	7.815E-11	6.872E-11	8.269E-11
	ReCiPe 2016 Midpoint (E) - Human toxicity, cancer	2.419E-04	2.956E-04	2.466E-04	2.549E-04	2.640E-04	2.274E-04	2.666E-04	2.760E-04	2.391E-04	2.723E-04	2.845E-04	2.845E-04	2.597E-04	2.720E-04	2.230E-04	2.715E-04	2.842E-04	2.347E-04	2.487E-04	2.610E-04	2.120E-04	2.604E-04
	TRACI 2.1, Human toxicity, cancer	2.843E-11	3.245E-11	2.847E-11	3.233E-11	3.341E-11	3.043E-11	3.305E-11	3.414E-11	3.116E-11	3.079E-11	3.224E-11	3.224E-11	2.961E-11	3.106E-11	2.708E-11	3.034E-11	3.179E-11	2.781E-11	2.941E-11	3.086E-11	2.688E-11	3.013E-11
Photochem. Ozone Creation Potential (POCP)	CML2001 - Jan. 2016, Photochem. Ozone Creation Potential (POCP)	6.245E-06	7.989E-06	5.689E-06	9.182E-06	1.055E-05	8.822E-06	8.588E-06	9.953E-06	8.228E-06	6.148E-06	7.968E-06	7.968E-06	7.332E-06	9.152E-06	6.852E-06	6.738E-06	8.561E-06	5.651E-06	7.312E-06	9.131E-06	6.832E-06	6.718E-06
	I02+ v2.1 - Photochemical oxidation - Midpoint	4.810E-06	5.820E-06	4.845E-06	5.687E-06	5.847E-06	5.116E-06	5.819E-06	5.981E-06	5.248E-06	5.589E-06	5.803E-06	5.803E-06	5.342E-06	5.556E-06	4.580E-06	5.474E-06	5.691E-06	4.713E-06	5.325E-06	5.539E-06	4.564E-06	5.457E-06
	ILCD - Photochemical ozone formation midpoint, human health (v1.09)	1.323E-04	1.543E-04	1.343E-04	1.543E-04	1.563E-04	1.413E-04	1.583E-04	1.604E-04	1.453E-04	1.514E-04	1.541E-04	1.541E-04	1.437E-04	1.464E-04	1.263E-04	1.477E-04	1.504E-04	1.303E-04	1.435E-04	1.462E-04	1.262E-04	1.475E-04
	ReCiPe 2016 Midpoint (H) - Photochemical Ozone Formation, Ecosystems	1.750E-02	4.376E-02	1.753E-02	2.962E-02	3.870E-02	1.903E-02	2.989E-02	3.898E-02	1.931E-02	3.154E-02	4.366E-02	4.366E-02	3.109E-02	4.321E-02	1.698E-02	3.137E-02	4.348E-02	1.725E-02	3.100E-02	4.311E-02	1.688E-02	3.127E-02
Ozone Layer Depletion Potential	CML2001 - Jan. 2016 Ozone Layer Depletion Potential (ODP, steady state)	3.570E-13	3.755E-13	3.570E-13	3.580E-13	3.717E-13	3.578E-13	3.583E-13	3.720E-13	3.582E-13	3.572E-13	3.754E-13	3.754E-13	3.566E-13	3.748E-13	3.564E-13	3.569E-13	3.752E-13	3.567E-13	3.565E-13	3.748E-13	3.563E-13	3.568E-13
	I02+ v2.1 - Ozone layer depletion - Midpoint	3.570E-13	3.755E-13	3.570E-13	3.580E-13	3.717E-13	3.578E-13	3.583E-13	3.720E-13	3.582E-13	3.572E-13	3.754E-13	3.570E-13	3.566E-13	3.748E-13	3.564E-13	3.569E-13	3.752E-13	3.567E-13	3.565E-13	3.748E-13	3.563E-13	3.568E-13
	ILCD - Ozone depletion midpoint (v1.09)	3.572E-13	3.760E-13	3.572E-13	3.582E-13	3.721E-13	3.580E-13	3.585E-13	3.724E-13	3.584E-13	3.573E-13	3.759E-13	3.759E-13	3.568E-13	3.753E-13	3.566E-13	3.571E-13	3.756E-13	3.569E-13	3.567E-13	3.752E-13	3.565E-13	3.570E-13
	ReCiPe 2016 Midpoint (H) - Stratospheric Ozone Depletion	1.312E-08	2.377E-08	1.325E-08	2.038E-08	2.180E-08	1.390E-08	2.064E-08	2.206E-08	1.417E-08	2.177E-08	2.366E-08	2.366E-08	2.140E-08	2.328E-08	1.276E-08	2.166E-08	2.355E-08	1.302E-08	2.128E-08	2.317E-08	1.265E-08	2.155E-08
	TRACI 2.1, Ozone Depletion Air	3.790E-13	3.974E-13	3.790E-13	3.799E-13	3.936E-13	3.798E-13	3.803E-13	3.940E-13	3.801E-13	3.791E-13	3.974E-13	3.974E-13	3.785E-13	3.968E-13	3.783E-13	3.789E-13	3.971E-13	3.787E-13	3.785E-13	3.967E-13	3.783E-13	3.788E-13
Terrestrial Acidification	CML2001 - Jan. 2016, Acidification Potential (AP)	1.256E-04	1.500E-04	1.270E-04	1.515E-04	1.533E-04	1.361E-04	1.543E-04	1.561E-04	1.389E-04	1.467E-04	1.491E-04	1.491E-04	1.419E-04	1.443E-04	1.214E-04	1.448E-04	1.472E-04	1.243E-04	1.411E-04	1.435E-04	1.205E-04	1.439E-04
	I02+ v2.1 - Terrestrial acidification/nutrification - Midpoint	7.066E-04	8.292E-04	7.213E-04	8.199E-04	8.226E-04	7.416E-04	8.462E-04	8.489E-04	7.680E-04	8.247E-04	8.283E-04	8.283E-04	7.738E-04	7.774E-04	6.695E-04	8.001E-04	8.038E-04	6.958E-04	7.729E-04	7.765E-04	6.686E-04	7.992E-04
	ILCD Acidification midpoint (v1.09)	1.557E-04	1.841E-04	1.591E-04	1.893E-04	1.886E-04	1.699E-04	1.933E-04	1.927E-04	1.740E-04	1.847E-04	1.838E-04	1.838E-04	1.770E-04	1.761E-04	1.511E-04	1.810E-04	1.802E-04	1.552E-04	1.767E-04	1.759E-04	1.509E-04	1.808E-04
	ReCiPe 2016 Midpoint (E) - Terrestrial Acidification	9.559E-05	1.121E-04	9.665E-05	1.166E-04	1.160E-04	1.044E-04	1.187E-04	1.181E-04	1.065E-04	1.127E-04	1.119E-04	1.119E-04	1.097E-04	1.076E-04	9.246E-05	1.108E-04	1.100E-04	9.459E-05	1.085E-04	1.077E-04	9.231E-05	1.107E-04
	TRACI 2.1, Acidification	1.434E-04	1.694E-04	1.454E-04	1.707E-04	1.722E-04	1.542E-04	1.745E-04	1.760E-04	1.580E-04	1.672E-04	1.692E-04	1.692E-04	1.600E-04	1.620E-04	1.380E-04	1.638E-04	1.658E-04	1.418E-04	1.598E-04	1.618E-04	1.378E-04	1.636E-04
Terrestrial ecotoxicity	CML2001 - Jan. 2016, Terrestric Ecotoxicity Potential (TETP inf.)	1.151E-04	1.243E-04	1.152E-04	1.351E-04	1.317E-04	1.249E-04	1.382E-04	1.348E-04	1.280E-04	1.287E-04	1.242E-04	1.242E-04	1.235E-04	1.190E-04	1.099E-04	1.266E-04	1.221E-04	1.130E-04	1.234E-04	1.189E-04	1.098E-04	1.265E-04
	ReCiPe 2016 Midpoint (H) - Terrestrial ecotoxicity	1.785E-02	2.246E-02	1.786E-02	2.106E-02	2.364E-02	2.018E-02	2.123E-02	2.381E-02	2.036E-02	1.900E-02	2.245E-02	2.245E-02	1.868E-02	2.212E-02	1.751E-02	1.885E-02	2.230E-02	1.769E-02	1.866E-02	2.211E-02	1.750E-02	1.883E-02

Table 4.11 Life-cycle greenhouse gas emissions of 1 kg 40 MPa concrete

Impact category	Life-cycle assessment method	Unit	0-0-0	0-1-0	0-0-1	0-0-2	1-0-0	2-0-0	3-0-0	1-1-0	2-1-0	3-1-0	1-0-1	2-0-1	3-0-1	1-0-2	2-0-2	3-0-2
	Concrete mixture option		(1)0-0	(2)0-0.5W	(3)0-0.5Cs	(4)0-0.5S	(5)1F-0	(7)1S-0	(6)1B-0	(8)1F-0.5W	(10)1S-0.5W	(9)1B-0.5W	(11)1F-0.5Cs	(13)1S-0.5Cs	(12)1B-0.5Cs	(14)1F-0.5S	(16)1S-0.5S	(15)1B-0.5S
	Points to achieve			0.5	0.5	0.5	1	1	1	1.5	1.5	1.5	1.5	1.5	1.5	1.5	1.5	1.5
Global Warming Potential (GWP 100 years)	CML2001 - Jan. 2016, Global Warming Potential (GWP 100 years)	kg CO2 eq.	0.1509	0.1510	0.1498	0.1230	0.1195	0.1230	0.1088	0.1193	0.1228	0.1086	0.1181	0.1216	0.1074	0.1189	0.1224	0.1082
	I02+ v2.1 - Global warming 500yr - Midpoint	kg CO2 eq.	0.1505	0.1505	0.1492	0.1227	0.1182	0.1216	0.1081	0.1182	0.1215	0.1081	0.1169	0.1202	0.1071	0.1176	0.1210	0.1079
	ILCD - Climate change midpoint, incl biogenic carbon (v1.09)	kg CO2 eq.	0.1508	0.1508	0.1494	0.1230	0.1190	0.1224	0.1084	0.1190	0.1223	0.1083	0.1176	0.1210	0.1070	0.1184	0.1218	0.1078
	ReCiPe 2016 Midpoint (H) - Climate change, incl biogenic carbon	kg CO2 eq.	0.1496	0.1494	0.1481	0.1218	0.1066	0.1207	0.1073	0.1173	0.1207	0.1073	0.1160	0.1194	0.1060	0.1168	0.1201	0.1068
	TRACI 2.1, Global Warming Air, incl. biogenic carbon	kg CO2 eq.	0.1508	0.1507	0.1494	0.1229	0.1075	0.1223	0.1084	0.1189	0.1222	0.1083	0.1176	0.1209	0.1070	0.1184	0.1217	0.1078
Aquatic Ecotoxicity	CML2001 - Jan. 2016, Freshwater Aquatic Ecotoxicity Pot. (FAETP inf.)	kg DCB eq.	1.203E-04	1.198E-04	1.169E-04	9.476E-05	1.008E-04	1.061E-04	8.995E-05	9.977E-05	1.055E-04	8.943E-05	9.686E-05	1.026E-04	8.653E-05	9.882E-05	1.046E-04	8.848E-05
	I02+ v2.1 - Aquatic ecotoxicity - Midpoint	kg TEG eq. to water	3.039E-01	3.015E-01	2.873E-01	2.304E-01	3.142E-01	3.577E-01	2.421E-01	3.118E-01	3.553E-01	2.397E-01	2.977E-01	3.412E-01	2.256E-01	3.081E-01	3.517E-01	2.360E-01
	ILCD Ecotoxicity freshwater midpoint (v1.09)	CTUe	2.558E-03	2.536E-03	2.433E-03	1.985E-03	2.382E-03	2.095E-03	2.002E-03	2.360E-03	2.073E-03	1.979E-03	2.257E-03	1.970E-03	1.876E-03	2.331E-03	2.046E-03	1.950E-03
	ReCiPe 2016 Midpoint (E) - Freshwater ecotoxicity	kg 1,4 DB eq.	4.850E-06	4.808E-06	4.689E-06	3.791E-06	3.658E-06	4.477E-06	3.687E-06	4.276E-06	4.435E-06	3.644E-06	4.157E-06	4.316E-06	3.526E-06	4.248E-06	4.409E-06	3.617E-06
	TRACI 2.1, Ecotoxicity	CTUe	1.451E-03	1.439E-03	1.387E-03	1.105E-03	1.110E-03	1.102E-03	1.118E-03	1.405E-03	1.090E-03	1.106E-03	1.353E-03	1.039E-03	1.055E-03	1.394E-03	1.080E-03	1.096E-03
Eutrophication Potential (EP)	CML2001 - Jan. 2016, Eutrophication Potential (EP)	kg PO4 eq.	2.798E-05	2.812E-05	2.692E-05	2.259E-05	2.508E-05	2.598E-05	2.244E-05	2.487E-05	2.577E-05	2.223E-05	2.367E-05	2.456E-05	2.109E-05	2.448E-05	2.538E-05	2.184E-05
	I02+ v2.1 - Aquatic eutrophication - Midpoint	kg PO4 eq.	2.670E-07	2.577E-07	2.536E-07	1.811E-07	2.162E-07	3.663E-07	2.128E-07	2.069E-07	3.571E-07	2.036E-07	2.028E-07	3.529E-07	1.994E-07	2.108E-07	3.610E-07	2.075E-07
	ILCD Eutrophication freshwater midpoint (v1.09)	kg PO4 eq.	8.603E-08	8.475E-08	8.338E-08	5.957E-08	7.105E-08	1.205E-07	6.994E-08	6.802E-08	1.175E-07	6.691E-08	6.665E-08	1.161E-07	6.554E-08	6.930E-08	1.187E-07	6.819E-08
	ReCiPe 2016 Midpoint (H) - Freshwater Eutrophication	kg PO4 eq.	8.770E-08	8.468E-08	8.330E-08	5.951E-08	6.983E-08	1.204E-07	6.988E-08	6.795E-08	1.173E-07	6.685E-08	6.658E-08	1.160E-07	6.548E-08	6.923E-08	1.186E-07	6.813E-08
	TRACI 2.1, Eutrophication	kg N eq.	1.409E-05	1.404E-05	1.352E-05	1.197E-05	1.113E-05	1.445E-05	1.142E-05	1.325E-05	1.440E-05	1.137E-05	1.273E-05	1.388E-05	1.085E-05	1.306E-05	1.421E-05	1.119E-05
Human toxicity	CML2001 - Jan. 2016, Human Toxicity Potential (HTP inf.)	kg DCB eq.	3.286E-02	3.285E-02	3.281E-02	2.666E-02	2.344E-02	2.374E-02	2.308E-02	2.343E-02	2.374E-02	2.307E-02	2.339E-02	2.370E-02	2.303E-02	2.342E-02	2.372E-02	2.305E-02
	ILCD Human toxicity midpoint, cancer effects (v1.09)	CTUh	1.124E-10	1.112E-10	1.052E-10	8.726E-11	1.002E-10	9.925E-11	9.090E-11	9.894E-11	9.804E-11	8.968E-11	9.302E-11	9.211E-11	8.376E-11	9.686E-11	9.608E-11	8.760E-11
	ReCiPe 2016 Midpoint (E) - Human toxicity, cancer	kg 1,4-DB eq.	3.359E-04	3.249E-04	3.124E-04	2.424E-04	2.767E-04	3.245E-04	2.809E-04	3.027E-04	3.135E-04	2.699E-04	2.902E-04	3.010E-04	2.574E-04	3.028E-04	3.139E-04	2.700E-04
	TRACI 2.1, Human toxicity, cancer	CTUh	4.925E-11	4.905E-11	4.787E-11	3.739E-11	3.653E-11	4.011E-11	3.656E-11	3.862E-11	3.991E-11	3.636E-11	3.744E-11	3.873E-11	3.518E-11	3.823E-11	3.950E-11	3.595E-11
Ozone Creation Potential (POCP)	CML2001 - Jan. 2016, Photochem. Ozone Creation Potential (POCP)	kg C2H4 eq. to air	1.675E-05	1.679E-05	1.819E-05	1.305E-05	9.991E-06	1.161E-05	9.562E-06	9.970E-06	1.159E-05	9.542E-06	1.142E-05	1.305E-05	1.100E-05	1.057E-05	1.219E-05	1.014E-05
	I02+ v2.1 - Photochemical oxidation - Midpoint	kg C2H4 eq. to air	8.068E-06	8.072E-06	7.821E-06	6.310E-06	6.798E-06	6.988E-06	6.118E-06	6.781E-06	6.971E-06	6.101E-06	6.530E-06	6.720E-06	5.850E-06	6.675E-06	6.867E-06	5.995E-06
	ILCD - Photochemical ozone formation midpoint, human health (v1.09)	kg NMVOC eq.	2.254E-04	2.252E-04	2.166E-04	1.788E-04	1.667E-04	1.893E-04	1.715E-04	1.868E-04	1.892E-04	1.713E-04	1.782E-04	1.806E-04	1.627E-04	1.832E-04	1.856E-04	1.677E-04
	ReCiPe 2016 Midpoint (H) - Photochemical Ozone Formation, Ecosystems	kg NOx eq.	3.016E-02	3.006E-02	2.961E-02	2.418E-02	2.245E-02	4.588E-02	2.247E-02	3.497E-02	4.579E-02	2.238E-02	3.452E-02	4.534E-02	2.193E-02	3.482E-02	4.563E-02	2.222E-02
Ozone Layer Depletion Potential	CML2001 - Jan. 2016 Ozone Layer Depletion Potential (ODP, steady state)	kg R-11 eq.	3.660E-13	3.659E-13	3.653E-13	4.384E-13	3.607E-13	3.770E-13	3.605E-13	3.606E-13	3.769E-13	3.604E-13	3.600E-13	3.763E-13	3.599E-13	3.604E-13	3.767E-13	3.602E-13
	I02+ v2.1 - Ozone layer depletion - Midpoint	kg R-11 eq.	3.660E-13	3.658E-13	3.653E-13	4.384E-13	3.607E-13	3.770E-13	3.605E-13	3.606E-13	3.769E-13	3.604E-13	3.600E-13	3.763E-13	3.599E-13	3.604E-13	3.767E-13	3.602E-13
	ILCD - Ozone depletion midpoint (v1.09)	kg CFC-11 eq.	3.663E-13	3.662E-13	3.657E-13	4.386E-13	3.609E-13	3.775E-13	3.607E-13	3.608E-13	3.774E-13	3.607E-13	3.603E-13	3.768E-13	3.601E-13	3.606E-13	3.772E-13	3.604E-13
	ReCiPe 2016 Midpoint (H) - Stratospheric Ozone Depletion	kg CFC-11 eq.	2.063E-08	2.065E-08	2.009E-08	1.762E-08	1.599E-08	2.569E-08	1.629E-08	2.389E-08	2.558E-08	1.618E-08	2.347E-08	2.515E-08	1.575E-08	2.379E-08	2.547E-08	1.608E-08
	TRACI 2.1, Ozone Depletion Air	kg CFC-11 eq.	3.879E-13	3.879E-13	3.873E-13	4.658E-13	3.825E-13	3.989E-13	3.825E-13	3.862E-11	3.989E-13	3.824E-13	3.820E-13	3.983E-13	3.818E-13	3.823E-13	3.986E-13	3.822E-13
Terrestrial Acidification	CML2001 - Jan. 2016, Acidification Potential (AP)	kg SO2 eq.	2.199E-04	2.190E-04	2.136E-04	1.752E-04	1.829E-04	1.850E-04	1.646E-04	1.820E-04	1.842E-04	1.637E-04	1.767E-04	1.788E-04	1.583E-04	1.802E-04	1.823E-04	1.619E-04
	I02+ v2.1 - Terrestrial acidification/nutrification - Midpoint	kg SO2 eq. to air	1.166E-03	1.165E-03	1.108E-03	9.274E-04	1.000E-03	1.003E-03	9.070E-04	9.992E-04	1.002E-03	9.061E-04	9.419E-04	9.450E-04	8.487E-04	9.758E-04	9.785E-04	8.822E-04
	ILCD Acidification midpoint (v1.09)	Mole of H+ eq.	2.730E-04	2.727E-04	2.641E-04	2.186E-04	2.284E-04	2.276E-04	2.054E-04	2.282E-04	2.274E-04	2.051E-04	2.196E-04	2.188E-04	1.966E-04	2.246E-04	2.339E-04	2.016E-04
	ReCiPe 2016 Midpoint (E) - Terrestrial Acidification	kg SO2 eq.	1.688E-04	1.686E-04	1.641E-04	1.354E-04	1.233E-04	1.395E-04	1.257E-04	1.401E-04	1.394E-04	1.256E-04	1.356E-04	1.349E-04	1.211E-04	1.383E-04	1.375E-04	1.238E-04
	TRACI 2.1, Acidification	kg SO2 eq.	2.440E-04	2.437E-04	2.358E-04	1.974E-04	1.809E-04	2.070E-04	1.856E-04	2.050E-04	2.067E-04	1.854E-04	1.970E-04	1.988E-04	1.774E-04	2.017E-04	2.035E-04	1.821E-04
Terrestrial ecotoxicity	CML2001 - Jan. 2016, Terrestric Ecotoxicity Potential (TETP inf.)	kg DCB eq.	2.036E-04	2.035E-04	1.983E-04	1.558E-04	1.618E-04	1.578E-04	1.497E-04	1.617E-04	1.577E-04	1.496E-04	1.565E-04	1.525E-04	1.444E-04	1.598E-04	1.558E-04	1.476E-04
	ReCiPe 2016 Midpoint (H) - Terrestrial ecotoxicity	kg 1,4-DB eq.	3.367E-02	3.365E-02	3.332E-02	2.677E-02	2.406E-02	2.818E-02	2.407E-02	2.509E-02	2.817E-02	2.405E-02	2.477E-02	2.784E-02	2.373E-02	2.495E-02	2.803E-02	2.391E-02

Life-cycle assessment method	4-0-0	5-0-0	6-0-0	1-1-1	2-1-1	3-1-1	1-1-2	2-1-2	3-1-2	4-1-0	5-1-0	6-1-0	4-0-1	5-0-1	6-0-1	4-0-2	5-0-2	6-0-2	4-1-1	5-1-1	6-1-1	4-1-2
Concrete mixture option	(17)2F-0	(19)2S-0	(18)2B-0	(29)1F-1Cs	(31)1S-1Cs	(30)1B-1Cs	(32)1F-1S	(34)1S-1S	(33)1B-1S	(20)2F-0.5W	(22)2S-0.5W	(21)2B-0.5W	(23)2F-0.5Cs	(25)2S-0.5Cs	(24)2B-0.5Cs	(26)2F-0.5S	(28)2S-0.5S	(27)2B-0.5S	(35)2F-1Cs	(37)2S-1Cs	(36)2B-1Cs	(38)2F-1S
Points to achieve	2	2	2	2	2	2	2	2	2	2.5	2.5	2.5	2.5	2.5	2.5	2.5	2.5	2.5	3	3	3	3
CML2001 - Jan. 2016, Global Warming Potential (GWP 100 years)	0.0942	0.1136	0.0946	0.1179	0.1214	0.1072	0.1187	0.1222	0.1080	0.1087	0.1134	0.1134	0.1075	0.1122	0.0932	0.1083	0.1130	0.0940	0.1073	0.1120	0.0930	0.1081
I02+ v2.1 - Global warming 500yr - Midpoint	0.0936	0.1119	0.0943	0.1168	0.1202	0.1067	0.1176	0.1210	0.1075	0.1074	0.1119	0.1119	0.1061	0.1106	0.0930	0.1069	0.1114	0.0937	0.1060	0.1105	0.0926	0.1068
ILCD - Climate change midpoint, incl biogenic carbon (v1.09)	0.0938	0.1129	0.0943	0.1176	0.1209	0.1070	0.1184	0.1217	0.1077	0.1084	0.1128	0.1128	0.1071	0.1115	0.0929	0.1078	0.1123	0.0937	0.1070	0.1114	0.0928	0.1078
ReCiPe 2016 Midpoint (H) - Climate change, incl biogenic carbon	0.0929	0.1111	0.0933	0.1160	0.1193	0.1059	0.1167	0.1201	0.1067	0.1066	0.1111	0.1111	0.1053	0.1098	0.0920	0.1061	0.1106	0.0927	0.1053	0.1097	0.0919	0.1060
TRACI 2.1, Global Warming Air, incl. biogenic carbon	0.0938	0.1128	0.0942	0.1176	0.1208	0.1069	0.1183	0.1216	0.1077	0.1084	0.1127	0.1127	0.1070	0.1114	0.0928	0.1078	0.1122	0.0936	0.1070	0.1113	0.0928	0.1078
CML2001 - Jan. 2016, Freshwater Aquatic Ecotoxicity Pot. (FAETP inf.)	7.904E-05	1.013E-04	7.982E-05	9.634E-05	1.021E-04	8.601E-05	9.830E-05	1.041E-04	8.796E-05	9.308E-05	1.008E-04	1.008E-04	9.018E-05	9.792E-05	7.640E-05	9.213E-05	9.995E-05	7.835E-05	8.966E-05	9.740E-05	7.588E-05	9.161E-05
I02+ v2.1 - Aquatic ecotoxicity - Midpoint	2.193E-01	3.757E-01	2.216E-01	2.953E-01	3.388E-01	2.232E-01	3.057E-01	3.493E-01	2.336E-01	3.153E-01	3.733E-01	3.733E-01	3.011E-01	3.592E-01	2.050E-01	3.116E-01	3.699E-01	2.155E-01	2.987E-01	3.568E-01	2.026E-01	3.092E-01
ILCD Ecotoxicity freshwater midpoint (v1.09)	1.783E-03	1.942E-03	1.816E-03	2.234E-03	1.947E-03	1.854E-03	2.308E-03	2.023E-03	1.928E-03	2.301E-03	1.920E-03	1.920E-03	2.198E-03	1.817E-03	1.691E-03	2.272E-03	1.894E-03	1.765E-03	2.176E-03	1.794E-03	1.668E-03	2.250E-03
ReCiPe 2016 Midpoint (E) - Freshwater ecotoxicity	3.261E-06	4.354E-06	3.299E-06	4.115E-06	4.274E-06	3.484E-06	4.206E-06	4.367E-06	3.574E-06	4.099E-06	4.312E-06	4.312E-06	3.980E-06	4.191E-06	3.138E-06	4.071E-06	4.288E-06	3.229E-06	3.938E-06	4.151E-06	3.096E-06	4.029E-06
TRACI 2.1, Ecotoxicity	9.965E-04	9.866E-04	1.007E-03	1.341E-03	1.027E-03	1.043E-03	1.382E-03	1.069E-03	1.084E-03	1.394E-03	9.748E-04	9.748E-04	1.342E-03	9.231E-04	9.437E-04	1.383E-03	9.652E-04	9.847E-04	1.330E-03	9.112E-04	9.319E-04	1.371E-03
CML2001 - Jan. 2016, Eutrophication Potential (EP)	2.006E-05	2.519E-05	2.048E-05	2.346E-05	2.435E-05	2.081E-05	2.427E-05	2.517E-05	2.163E-05	2.379E-05	2.498E-05	2.498E-05	2.259E-05	2.378E-05	1.906E-05	2.340E-05	2.460E-05	1.988E-05	2.237E-05	2.357E-05	1.885E-05	2.319E-05
I02+ v2.1 - Aquatic eutrophication - Midpoint	1.945E-07	3.994E-07	1.948E-07	1.935E-07	3.436E-07	1.901E-07	2.016E-07	3.517E-07	1.982E-07	1.900E-07	3.902E-07	3.902E-07	1.858E-07	3.860E-07	1.813E-07	1.939E-07	3.941E-07	1.894E-07	1.766E-07	3.767E-07	1.721E-07	1.846E-07
ILCD Eutrophication freshwater midpoint (v1.09)	6.392E-08	1.314E-07	6.400E-08	6.362E-08	1.130E-07	6.251E-08	6.627E-08	1.157E-07	6.516E-08	6.244E-08	1.284E-07	1.284E-07	6.107E-08	1.270E-07	5.959E-08	6.372E-08	1.296E-07	6.224E-08	5.804E-08	1.240E-07	5.656E-08	6.069E-08
ReCiPe 2016 Midpoint (H) - Freshwater Eutrophication	6.387E-08	1.313E-07	6.394E-08	6.355E-08	1.129E-07	6.245E-08	6.620E-08	1.156E-07	6.510E-08	6.237E-08	1.282E-07	1.282E-07	6.100E-08	1.269E-07	5.954E-08	6.365E-08	1.295E-07	6.219E-08	5.797E-08	1.238E-07	5.651E-08	6.062E-08
TRACI 2.1, Eutrophication	1.039E-05	1.457E-05	1.053E-05	1.268E-05	1.383E-05	1.081E-05	1.301E-05	1.416E-05	1.114E-05	1.298E-05	1.452E-05	1.452E-05	1.246E-05	1.400E-05	9.965E-06	1.280E-05	1.433E-05	1.030E-05	1.242E-05	1.395E-05	9.916E-06	1.275E-05
CML2001 - Jan. 2016, Human Toxicity Potential (HTP inf.)	1.981E-02	2.070E-02	1.981E-02	2.339E-02	2.369E-02	2.303E-02	2.341E-02	2.372E-02	2.305E-02	2.029E-02	2.070E-02	2.070E-02	2.025E-02	2.066E-02	1.977E-02	2.028E-02	2.068E-02	1.979E-02	2.025E-02	2.065E-02	1.976E-02	2.027E-02
ILCD Human toxicity midpoint, cancer effects (v1.09)	8.149E-11	9.496E-11	8.374E-11	9.181E-11	9.090E-11	8.255E-11	9.564E-11	9.487E-11	8.638E-11	9.487E-11	9.375E-11	9.375E-11	8.895E-11	8.783E-11	7.660E-11	9.278E-11	9.190E-11	8.044E-11	8.773E-11	8.661E-11	7.539E-11	9.157E-11
ReCiPe 2016 Midpoint (E) - Human toxicity, cancer	2.570E-04	3.209E-04	2.626E-04	2.792E-04	2.900E-04	2.464E-04	2.918E-04	3.029E-04	2.590E-04	2.953E-04	3.099E-04	3.099E-04	2.828E-04	2.974E-04	2.391E-04	2.954E-04	3.105E-04	2.516E-04	2.718E-04	2.864E-04	2.281E-04	2.843E-04
TRACI 2.1, Human toxicity, cancer	3.228E-11	3.707E-11	3.233E-11	3.724E-11	3.853E-11	3.498E-11	3.800E-11	3.930E-11	3.575E-11	3.514E-11	3.686E-11	3.686E-11	3.396E-11	3.569E-11	3.095E-11	3.473E-11	3.646E-11	3.172E-11	3.376E-11	3.548E-11	3.075E-11	3.453E-11
CML2001 - Jan. 2016, Photochem. Ozone Creation Potential (POCP)	7.827E-06	9.902E-06	7.165E-06	1.140E-05	1.303E-05	1.098E-05	1.055E-05	1.217E-05	1.012E-05	7.715E-06	9.881E-06	9.881E-06	9.170E-06	1.134E-05	8.599E-06	8.313E-06	1.048E-05	7.136E-06	9.149E-06	1.131E-05	8.578E-06	8.292E-06
I02+ v2.1 - Photochemical oxidation - Midpoint	5.421E-06	6.623E-06	5.461E-06	6.513E-06	6.704E-06	5.833E-06	6.658E-06	6.850E-06	5.978E-06	6.351E-06	6.606E-06	6.606E-06	6.100E-06	6.355E-06	5.193E-06	6.245E-06	6.503E-06	5.338E-06	6.083E-06	6.338E-06	5.177E-06	6.228E-06
ILCD - Photochemical ozone formation midpoint, human health (v1.09)	1.512E-04	1.773E-04	1.535E-04	1.780E-04	1.804E-04	1.625E-04	1.830E-04	1.854E-04	1.675E-04	1.740E-04	1.772E-04	1.772E-04	1.654E-04	1.686E-04	1.447E-04	1.703E-04	1.736E-04	1.497E-04	1.652E-04	1.684E-04	1.446E-04	1.702E-04
ReCiPe 2016 Midpoint (H) - Photochemical Ozone Formation, Ecosystems	1.988E-02	5.113E-02	1.991E-02	3.443E-02	4.524E-02	2.183E-02	3.472E-02	4.553E-02	2.212E-02	3.661E-02	5.103E-02	5.103E-02	3.616E-02	5.058E-02	1.936E-02	3.646E-02	5.088E-02	1.966E-02	3.606E-02	5.048E-02	1.927E-02	3.636E-02
CML2001 - Jan. 2016 Ozone Layer Depletion Potential (ODP, steady state)	3.587E-13	3.807E-13	3.587E-13	3.600E-13	3.763E-13	3.598E-13	3.603E-13	3.766E-13	3.601E-13	3.588E-13	3.806E-13	3.806E-13	3.583E-13	3.800E-13	3.580E-13	3.586E-13	3.804E-13	3.584E-13	3.582E-13	3.799E-13	3.580E-13	3.585E-13
I02+ v2.1 - Ozone layer depletion - Midpoint	3.587E-13	3.807E-13	3.587E-13	3.600E-13	3.763E-13	3.598E-13	3.603E-13	3.766E-13	3.601E-13	3.588E-13	3.806E-13	3.586E-13	3.583E-13	3.800E-13	3.580E-13	3.586E-13	3.804E-13	3.584E-13	3.582E-13	3.799E-13	3.580E-13	3.585E-13
ILCD - Ozone depletion midpoint (v1.09)	3.589E-13	3.812E-13	3.589E-13	3.602E-13	3.767E-13	3.600E-13	3.605E-13	3.771E-13	3.604E-13	3.590E-13	3.811E-13	3.811E-13	3.585E-13	3.806E-13	3.582E-13	3.588E-13	3.809E-13	3.586E-13	3.584E-13	3.805E-13	3.582E-13	3.587E-13
ReCiPe 2016 Midpoint (H) - Stratospheric Ozone Depletion	1.465E-08	2.733E-08	1.480E-08	2.335E-08	2.504E-08	1.564E-08	2.368E-08	2.536E-08	1.596E-08	2.497E-08	2.722E-08	2.722E-08	2.455E-08	2.679E-08	1.427E-08	2.487E-08	2.712E-08	1.459E-08	2.443E-08	2.668E-08	1.415E-08	2.476E-08
TRACI 2.1, Ozone Depletion Air	3.806E-13	4.026E-13	3.806E-13	3.819E-13	3.982E-13	3.817E-13	3.823E-13	3.985E-13	3.821E-13	3.808E-13	4.025E-13	4.025E-13	3.802E-13	4.020E-13	3.800E-13	3.806E-13	4.023E-13	3.803E-13	3.801E-13	4.019E-13	3.799E-13	3.805E-13
CML2001 - Jan. 2016, Acidification Potential (AP)	1.444E-04	1.734E-04	1.461E-04	1.758E-04	1.779E-04	1.575E-04	1.793E-04	1.815E-04	1.610E-04	1.697E-04	1.726E-04	1.726E-04	1.643E-04	1.672E-04	1.399E-04	1.679E-04	1.707E-04	1.434E-04	1.635E-04	1.663E-04	1.390E-04	1.670E-04
I02+ v2.1 - Terrestrial acidification/nutrification - Midpoint	8.032E-04	9.491E-04	8.207E-04	9.410E-04	9.441E-04	8.478E-04	9.744E-04	9.776E-04	8.813E-04	9.440E-04	9.482E-04	9.482E-04	8.866E-04	8.909E-04	7.625E-04	9.201E-04	9.244E-04	7.959E-04	8.857E-04	8.900E-04	7.616E-04	9.192E-04
ILCD Acidification midpoint (v1.09)	1.787E-04	2.126E-04	1.828E-04	2.194E-04	2.186E-04	1.963E-04	2.244E-04	2.236E-04	2.013E-04	2.133E-04	2.123E-04	2.123E-04	2.047E-04	2.037E-04	1.740E-04	2.098E-04	2.088E-04	1.790E-04	2.045E-04	2.035E-04	1.738E-04	2.095E-04
ReCiPe 2016 Midpoint (E) - Terrestrial Acidification	1.102E-04	1.298E-04	1.114E-04	1.355E-04	1.348E-04	1.210E-04	1.381E-04	1.374E-04	1.236E-04	1.306E-04	1.296E-04	1.296E-04	1.271E-04	1.248E-04	1.068E-04	1.287E-04	1.278E-04	1.094E-04	1.260E-04	1.250E-04	1.067E-04	1.286E-04
TRACI 2.1, Acidification	1.638E-04	1.946E-04	1.661E-04	1.968E-04	1.986E-04	1.772E-04	2.015E-04	2.032E-04	1.818E-04	1.921E-04	1.944E-04	1.944E-04	1.841E-04	1.865E-04	1.579E-04	1.888E-04	1.911E-04	1.626E-04	1.839E-04	1.862E-04	1.577E-04	1.886E-04
CML2001 - Jan. 2016, Terrestric Ecotoxicity Potential (TETP inf.)	1.317E-04	1.426E-04	1.317E-04	1.964E-04	1.525E-04	1.443E-04	1.597E-04	1.557E-04	1.476E-04	1.478E-04	1.425E-04	1.425E-04	1.426E-04	1.373E-04	1.264E-04	1.459E-04	1.406E-04	1.297E-04	1.425E-04	1.372E-04	1.263E-04	1.458E-04
ReCiPe 2016 Midpoint (H) - Terrestrial ecotoxicity	2.086E-02	2.636E-02	2.087E-02	2.475E-02	2.783E-02	2.371E-02	2.493E-02	2.801E-02	2.390E-02	2.224E-02	2.634E-02	2.634E-02	2.192E-02	2.602E-02	2.053E-02	2.210E-02	2.620E-02	2.072E-02	2.190E-02	2.600E-02	2.051E-02	2.208E-02

Table 4.12 Life-cycle greenhouse gas emissions of 1 kg 50 MPa concrete

Impact category	Life-cycle assessment method	Unit	0-0-0	0-1-0	0-0-1	0-0-2	1-0-0	2-0-0	3-0-0	1-1-0	2-1-0	3-1-0	1-0-1	2-0-1	3-0-1	1-0-2	2-0-2	3-0-2
	Concrete mixture option		(1)0-0	(2)0-0.5W	(3)0-0.5Cs	(4)0-0.5S	(5)1F-0	(7)1S-0	(6)1B-0	(8)1F-0.5W	(10)1S-0.5W	(9)1B-0.5W	(11)1F-0.5Cs	(13)1S-0.5Cs	(12)1B-0.5Cs	(14)1F-0.5S	(16)1S-0.5S	(15)1B-0.5S
	Points to achieve			0.5	0.5	0.5	1	1	1	1.5	1.5	1.5	1.5	1.5	1.5	1.5	1.5	1.5
Global Warming Potential (GWP 100 years)	CML2001 - Jan. 2016, Global Warming Potential (GWP 100 years)	kg CO_2 eq.	0.1589	0.1590	0.1576	0.1045	0.1259	0.1296	0.1146	0.1256	0.1294	0.1144	0.1242	0.1279	0.1129	0.1252	0.1289	0.1139
	I02+ v2.1 - Global warming 500yr - Midpoint	kg CO_2 eq.	0.1586	0.1585	0.1570	0.1041	0.1246	0.1282	0.1140	0.1246	0.1281	0.1139	0.1231	0.1266	0.1127	0.1240	0.1275	0.1136
	ILCD - Climate change midpoint, incl biogenic carbon (v1.09)	kg CO_2 eq.	0.1588	0.1588	0.1572	0.1044	0.1254	0.1289	0.1142	0.1254	0.1289	0.1142	0.1238	0.1273	0.1126	0.1248	0.1283	0.1136
	ReCiPe 2016 Midpoint (H) - Climate change, incl biogenic carbon	kg CO_2 eq.	0.1574	0.1574	0.1559	0.1034	0.1122	0.1272	0.1131	0.1237	0.1272	0.1131	0.1221	0.1257	0.1116	0.1231	0.1266	0.1125
	TRACI 2.1, Global Warming Air, incl. biogenic carbon	kg CO_2 eq.	0.1588	0.1587	0.1572	0.1044	0.1132	0.1289	0.1142	0.1254	0.1288	0.1141	0.1238	0.1272	0.1126	0.1248	0.1282	0.1136
Aquatic Ecotoxicity	CML2001 - Jan. 2016, Freshwater Aquatic Ecotoxicity Pot. (FAETP inf.)	kg DCB eq.	1.260E-04	1.255E-04	1.212E-04	8.325E-05	1.048E-04	1.109E-04	9.395E-05	1.044E-04	1.105E-04	9.351E-05	1.001E-04	1.062E-04	8.922E-05	1.041E-04	1.103E-04	9.324E-05
	I02+ v2.1 - Aquatic ecotoxicity - Midpoint	kg TEG eq. to water	3.177E-01	3.157E-01	2.948E-01	2.117E-01	3.286E-01	3.744E-01	2.527E-01	3.266E-01	3.724E-01	2.507E-01	3.058E-01	3.516E-01	2.298E-01	3.257E-01	3.716E-01	2.497E-01
	ILCD Ecotoxicity freshwater midpoint (v1.09)	CTUe	2.669E-03	2.650E-03	2.496E-03	1.794E-03	2.484E-03	2.181E-03	2.083E-03	2.465E-03	2.162E-03	2.064E-03	2.311E-03	2.008E-03	1.910E-03	2.459E-03	2.159E-03	2.058E-03
	ReCiPe 2016 Midpoint (E) - Freshwater ecotoxicity	kg 1,4 DB eq.	5.060E-06	5.025E-06	4.838E-06	3.366E-06	3.804E-06	4.667E-06	3.834E-06	4.464E-06	4.631E-06	3.799E-06	4.277E-06	4.445E-06	3.612E-06	4.466E-06	4.635E-06	3.800E-06
	TRACI 2.1, Ecotoxicity	CTUe	1.522E-03	1.512E-03	1.434E-03	9.870E-04	1.163E-03	1.155E-03	1.172E-03	1.477E-03	1.145E-03	1.162E-03	1.399E-03	1.067E-03	1.084E-03	1.476E-03	1.145E-03	1.161E-03
Eutrophication Potential (EP).	CML2001 - Jan. 2016, Eutrophication Potential (EP)	kg PO_4 eq.	2.951E-05	2.962E-05	2.816E-05	1.950E-05	2.647E-05	2.742E-05	2.369E-05	2.630E-05	2.724E-05	2.351E-05	2.484E-05	2.578E-05	2.205E-05	2.583E-05	2.677E-05	2.304E-05
	I02+ v2.1 - Aquatic eutrophication - Midpoint	kg PO_4 eq.	2.756E-07	2.678E-07	2.571E-07	1.653E-07	2.221E-07	3.803E-07	2.186E-07	2.143E-07	3.725E-07	2.108E-07	2.036E-07	3.618E-07	2.000E-07	2.195E-07	3.777E-07	2.160E-07
	ILCD Eutrophication freshwater midpoint (v1.09)	kg PO_4 eq.	8.979E-08	8.809E-08	8.455E-08	5.438E-08	7.301E-08	1.251E-07	7.184E-08	7.046E-08	1.225E-07	6.929E-08	6.692E-08	1.190E-07	6.575E-08	7.216E-08	1.242E-07	7.099E-08
	ReCiPe 2016 Midpoint (H) - Freshwater Eutrophication	kg PO_4 eq.	9.056E-08	8.801E-08	8.447E-08	5.433E-08	7.172E-08	1.250E-07	7.178E-08	7.038E-08	1.224E-07	6.923E-08	6.685E-08	1.189E-07	6.569E-08	7.208E-08	1.241E-07	7.093E-08
	TRACI 2.1, Eutrophication	kg N eq.	1.474E-05	1.469E-05	1.406E-05	1.081E-05	1.161E-05	1.515E-05	1.196E-05	1.389E-05	1.510E-05	1.192E-05	1.326E-05	1.447E-05	1.128E-05	1.370E-05	1.491E-05	1.172E-05
Human toxicity	CML2001 - Jan. 2016, Human Toxicity Potential (HTP inf.)	kg DCB eq	3.461E-02	3.461E-02	3.455E-02	2.251E-02	2.469E-02	2.501E-02	2.430E-02	2.468E-02	2.500E-02	2.430E-02	2.462E-02	2.494E-02	2.424E-02	2.468E-02	2.500E-02	2.429E-02
	ILCD Human toxicity midpoint, cancer effects (v1.09)	CTUh	1.162E-10	1.152E-10	1.063E-10	8.089E-11	1.033E-10	1.023E-10	9.355E-11	1.023E-10	1.013E-10	9.252E-11	9.344E-11	9.248E-11	8.368E-11	1.017E-10	1.009E-10	9.194E-11
	ReCiPe 2016 Midpoint (E) - Human toxicity, cancer	kg 1,4-DB eq.	3.430E-04	3.337E-04	3.105E-04	2.300E-04	2.806E-04	3.309E-04	2.850E-04	3.103E-04	3.217E-04	2.758E-04	2.871E-04	2.985E-04	2.526E-04	3.143E-04	3.260E-04	2.797E-04
	TRACI 2.1, Human toxicity, cancer	CTUh	5.153E-11	5.135E-11	4.962E-11	3.275E-11	3.812E-11	4.190E-11	3.816E-11	4.036E-11	4.172E-11	3.798E-11	3.863E-11	3.999E-11	3.625E-11	4.024E-11	4.160E-11	3.786E-11
Photochem. Ozone Creation Potential (POCP)	CML2001 - Jan. 2016, Photochem. Ozone Creation Potential (POCP)	kg C_2H_4 eq. to air	1.724E-05	1.722E-05	1.866E-05	1.170E-05	9.935E-06	1.165E-05	9.484E-06	9.918E-06	1.163E-05	9.467E-06	1.136E-05	1.307E-05	1.091E-05	1.084E-05	1.255E-05	1.039E-05
	I02+ v2.1 - Photochemical oxidation - Midpoint	kg C_2H_4 eq. to air	8.454E-06	8.440E-06	8.093E-06	5.579E-06	7.097E-06	7.297E-06	6.380E-06	7.082E-06	7.283E-06	6.366E-06	6.736E-06	6.936E-06	6.019E-06	7.027E-06	7.229E-06	6.310E-06
	ILCD - Photochemical ozone formation midpoint, human health (v1.09)	kg NMVOC eq.	2.375E-04	2.374E-04	2.271E-04	1.530E-04	1.757E-04	2.001E-04	1.813E-04	1.974E-04	1.999E-04	1.811E-04	1.872E-04	1.897E-04	1.709E-04	1.938E-04	1.963E-04	1.775E-04
	ReCiPe 2016 Midpoint (H) - Photochemical Ozone Formation, Ecosystems	kg NO_x eq.	3.151E-02	3.143E-02	3.076E-02	2.132E-02	2.339E-02	4.808E-02	2.341E-02	3.661E-02	4.800E-02	2.333E-02	3.594E-02	4.733E-02	2.266E-02	3.656E-02	4.796E-02	2.329E-02
Ozone Layer Depletion Potential	CML2001 - Jan. 2016 Ozone Layer Depletion Potential (ODP, steady state)	kg R-11 eq.	3.670E-13	3.669E-13	3.661E-13	4.591E-13	3.614E-13	3.785E-13	3.612E-13	3.613E-13	3.785E-13	3.611E-13	3.605E-13	3.776E-13	3.603E-13	3.612E-13	3.784E-13	3.610E-13
	I02+ v2.1 - Ozone layer depletion - Midpoint	kg R-11 eq.	3.670E-13	3.669E-13	3.661E-13	4.591E-13	3.614E-13	3.785E-13	3.612E-13	3.613E-13	3.785E-13	3.611E-13	3.605E-13	3.776E-13	3.603E-13	3.612E-13	3.784E-13	3.610E-13
	ILCD - Ozone depletion midpoint (v1.09)	kg CFC-11 eq.	3.673E-13	3.672E-13	3.664E-13	4.593E-13	3.616E-13	3.791E-13	3.614E-13	3.615E-13	3.790E-13	3.613E-13	3.607E-13	3.782E-13	3.605E-13	3.615E-13	3.789E-13	3.613E-13
	ReCiPe 2016 Midpoint (H) - Stratospheric Ozone Depletion	kg CFC-11 eq.	2.161E-08	2.162E-08	2.097E-08	1.571E-08	1.669E-08	2.694E-08	1.704E-08	2.507E-08	2.685E-08	1.695E-08	2.453E-08	2.630E-08	1.640E-08	2.494E-08	2.672E-08	1.682E-08
	TRACI 2.1, Ozone Depletion Air	kg CFC-11 eq.	3.889E-13	3.888E-13	3.880E-13	4.875E-13	3.831E-13	4.005E-13	3.831E-13	4.036E-11	4.004E-13	3.831E-13	3.824E-13	3.996E-13	3.822E-13	3.832E-13	4.003E-13	3.830E-13
Terrestrial Acidification	CML2001 - Jan. 2016, Acidification Potential (AP)	kg SO_2 eq.	2.313E-04	2.306E-04	2.239E-04	1.498E-04	1.927E-04	1.950E-04	1.734E-04	1.920E-04	1.942E-04	1.727E-04	1.853E-04	1.876E-04	1.660E-04	1.901E-04	1.923E-04	1.707E-04
	I02+ v2.1 - Terrestrial acidification/nutrification - Midpoint	kg SO_2 eq. to air	1.232E-03	1.231E-03	1.164E-03	7.898E-04	1.061E-03	1.064E-03	9.630E-04	1.060E-03	1.064E-03	9.622E-04	9.939E-04	9.973E-04	8.958E-04	1.034E-03	1.038E-03	9.360E-04
	ILCD Acidification midpoint (v1.09)	Mole of H+ eq.	2.877E-04	2.875E-04	2.773E-04	1.867E-04	2.413E-04	2.405E-04	2.170E-04	2.411E-04	2.403E-04	2.168E-04	2.309E-04	2.301E-04	2.066E-04	2.374E-04	2.366E-04	2.132E-04
	ReCiPe 2016 Midpoint (E) - Terrestrial Acidification	kg SO_2 eq.	1.777E-04	1.776E-04	1.722E-04	1.157E-04	1.298E-04	1.472E-04	1.327E-04	1.478E-04	1.471E-04	1.326E-04	1.425E-04	1.417E-04	1.272E-04	1.460E-04	1.453E-04	1.307E-04
	TRACI 2.1, Acidification	kg SO_2 eq.	2.569E-04	2.567E-04	2.472E-04	1.699E-04	1.904E-04	2.184E-04	1.959E-04	2.164E-04	2.182E-04	1.957E-04	2.069E-04	2.088E-04	1.862E-04	2.129E-04	2.148E-04	1.923E-04
Terrestrial ecotoxicity	CML2001 - Jan. 2016, Terrestric Ecotoxicity Potential (TETP inf.)	kg DCB eq.	2.141E-04	2.140E-04	2.068E-04	1.355E-04	1.701E-04	1.659E-04	1.573E-04	1.700E-04	1.658E-04	1.572E-04	1.628E-04	1.586E-04	1.500E-04	1.691E-04	1.649E-04	1.563E-04
	ReCiPe 2016 Midpoint (H) - Terrestrial ecotoxicity	kg 1,4-DB eq.	3.538E-02	3.537E-02	3.491E-02	2.288E-02	2.527E-02	2.961E-02	2.527E-02	2.635E-02	2.959E-02	2.526E-02	2.590E-02	2.914E-02	2.480E-02	2.629E-02	2.953E-02	2.520E-02

4-0-0	5-0-0	6-0-0	1-1-1	2-1-1	3-1-1	1-1-2	2-1-2	3-1-2	4-1-0	5-1-0	6-1-0	4-0-1	5-0-1	6-0-1	4-0-2	5-0-2	6-0-2	4-1-1	5-1-1	6-1-1	4-1-2
(17)2F-0	(19)2S-0	(18)2B-0	(29)1F-1Cs	(31)1S-1Cs	(30)1B-1Cs	(32)1F-1S	(34)1S-1S	(33)1B-1S	(20)2F-0.5W	(22)2S-0.5W	(21)2B-0.5W	(23)2F-0.5Cs	(25)2S-0.5Cs	(24)2B-0.5Cs	(26)2F-0.5S	(28)2S-0.5S	(27)2B-0.5S	(35)2F-1Cs	(37)2S-1Cs	(36)2B-1Cs	(38)2F-1S
2	2	2	2	2	2	2	2	2	2.5	2.5	2.5	2.5	2.5	2.5	2.5	2.5	2.5	3	3	3	3
0.0993	0.1197	0.0997	0.1240	0.1278	0.1127	0.1250	0.1288	0.1137	0.1145	0.1195	0.1195	0.1131	0.1181	0.0981	0.1141	0.1191	0.0991	0.1129	0.1179	0.0979	0.1139
0.0987	0.1180	0.0994	0.1230	0.1266	0.1124	0.1239	0.1275	0.1133	0.1133	0.1180	0.1180	0.1117	0.1165	0.0978	0.1127	0.1174	0.0988	0.1117	0.1164	0.0975	0.1126
0.0989	0.1190	0.0994	0.1238	0.1273	0.1126	0.1248	0.1283	0.1136	0.1143	0.1189	0.1189	0.1127	0.1173	0.0977	0.1137	0.1183	0.0987	0.1126	0.1173	0.0977	0.1136
0.0980	0.1172	0.0984	0.1221	0.1256	0.1115	0.1230	0.1266	0.1125	0.1124	0.1171	0.1171	0.1109	0.1156	0.0968	0.1118	0.1165	0.0978	0.1108	0.1155	0.0968	0.1118
0.0989	0.1189	0.0993	0.1237	0.1272	0.1125	0.1247	0.1282	0.1135	0.1142	0.1188	0.1188	0.1127	0.1173	0.0977	0.1136	0.1183	0.0987	0.1126	0.1172	0.0977	0.1136
8.246E-05	1.060E-04	8.328E-05	9.967E-05	1.058E-04	8.878E-05	1.037E-04	1.098E-04	9.280E-05	9.736E-05	1.055E-04	1.055E-04	9.307E-05	1.012E-04	7.855E-05	9.708E-05	1.053E-04	8.256E-05	9.263E-05	1.008E-04	7.811E-05	9.665E-05
2.286E-01	3.935E-01	2.310E-01	3.037E-01	3.496E-01	2.278E-01	3.236E-01	3.696E-01	2.477E-01	3.303E-01	3.914E-01	3.914E-01	3.094E-01	3.706E-01	2.081E-01	3.293E-01	3.907E-01	2.280E-01	3.074E-01	3.686E-01	2.061E-01	3.273E-01
1.852E-03	2.020E-03	1.888E-03	2.292E-03	1.989E-03	1.891E-03	2.440E-03	2.140E-03	2.039E-03	2.403E-03	2.001E-03	2.001E-03	2.249E-03	1.847E-03	1.714E-03	2.397E-03	1.999E-03	1.863E-03	2.230E-03	1.828E-03	1.695E-03	2.378E-03
3.386E-06	4.537E-06	3.426E-06	4.242E-06	4.409E-06	3.576E-06	4.430E-06	4.600E-06	3.765E-06	4.277E-06	4.502E-06	4.502E-06	4.091E-06	4.313E-06	3.203E-06	4.279E-06	4.508E-06	3.392E-06	4.055E-06	4.279E-06	3.168E-06	4.243E-06
1.044E-03	1.033E-03	1.055E-03	1.389E-03	1.057E-03	1.074E-03	1.466E-03	1.135E-03	1.151E-03	1.465E-03	1.023E-03	1.023E-03	1.387E-03	9.455E-04	9.673E-04	1.464E-03	1.024E-03	1.044E-03	1.377E-03	9.355E-04	9.573E-04	1.454E-03
2.121E-05	2.663E-05	2.165E-05	2.466E-05	2.560E-05	2.187E-05	2.565E-05	2.659E-05	2.286E-05	2.519E-05	2.645E-05	2.645E-05	2.373E-05	2.499E-05	2.001E-05	2.472E-05	2.598E-05	2.101E-05	2.355E-05	2.481E-05	1.984E-05	2.454E-05
1.993E-07	4.152E-07	1.995E-07	1.957E-07	3.539E-07	1.922E-07	2.117E-07	3.699E-07	2.082E-07	1.965E-07	4.074E-07	4.074E-07	1.857E-07	3.967E-07	1.810E-07	2.017E-07	4.126E-07	1.969E-07	1.779E-07	3.888E-07	1.732E-07	1.939E-07
6.550E-08	1.366E-07	6.558E-08	6.437E-08	1.165E-07	6.320E-08	6.960E-08	1.217E-07	6.844E-08	6.458E-08	1.340E-07	1.340E-07	6.104E-08	1.305E-07	5.949E-08	6.628E-08	1.357E-07	6.472E-08	5.849E-08	1.279E-07	5.693E-08	6.373E-08
6.544E-08	1.365E-07	6.552E-08	6.430E-08	1.163E-07	6.314E-08	6.953E-08	1.216E-07	6.837E-08	6.451E-08	1.339E-07	1.339E-07	6.098E-08	1.304E-07	5.944E-08	6.621E-08	1.356E-07	6.467E-08	5.842E-08	1.278E-07	5.688E-08	6.366E-08
1.088E-05	1.528E-05	1.103E-05	1.321E-05	1.443E-05	1.124E-05	1.366E-05	1.487E-05	1.168E-05	1.362E-05	1.524E-05	1.524E-05	1.299E-05	1.461E-05	1.036E-05	1.343E-05	1.505E-05	1.080E-05	1.295E-05	1.457E-05	1.031E-05	1.339E-05
2.086E-02	2.181E-02	2.087E-02	2.462E-02	2.494E-02	2.424E-02	2.467E-02	2.499E-02	2.429E-02	2.137E-02	2.180E-02	2.180E-02	2.132E-02	2.174E-02	2.081E-02	2.137E-02	2.180E-02	2.086E-02	2.131E-02	2.174E-02	2.080E-02	2.136E-02
8.363E-11	9.783E-11	8.600E-11	9.241E-11	9.146E-11	8.266E-11	1.007E-10	9.986E-11	9.092E-11	9.799E-11	9.680E-11	9.680E-11	8.915E-11	8.796E-11	7.614E-11	9.741E-11	9.647E-11	8.440E-11	8.812E-11	8.694E-11	7.511E-11	9.638E-11
2.598E-04	3.272E-04	2.657E-04	2.779E-04	2.892E-04	2.433E-04	3.050E-04	3.167E-04	2.704E-04	3.025E-04	3.179E-04	3.179E-04	2.793E-04	2.947E-04	2.333E-04	3.065E-04	3.225E-04	2.604E-04	2.701E-04	2.854E-04	2.240E-04	2.972E-04
3.365E-11	3.869E-11	3.370E-11	3.845E-11	3.981E-11	3.608E-11	4.007E-11	4.143E-11	3.769E-11	3.670E-11	3.852E-11	3.852E-11	3.496E-11	3.678E-11	3.179E-11	3.657E-11	3.840E-11	3.340E-11	3.479E-11	3.661E-11	3.162E-11	3.640E-11
7.597E-06	9.784E-06	6.900E-06	1.135E-05	1.306E-05	1.089E-05	1.082E-05	1.253E-05	1.037E-05	7.485E-06	9.767E-06	9.767E-06	8.930E-06	1.121E-05	8.328E-06	8.404E-06	1.069E-05	7.292E-06	8.913E-06	1.119E-05	8.311E-06	8.387E-06
5.646E-06	6.913E-06	5.689E-06	6.721E-06	6.922E-06	6.005E-06	7.012E-06	7.215E-06	6.296E-06	6.630E-06	6.899E-06	6.899E-06	6.283E-06	6.552E-06	5.328E-06	6.574E-06	6.846E-06	5.619E-06	6.269E-06	6.538E-06	5.314E-06	6.560E-06
1.600E-04	1.876E-04	1.625E-04	1.870E-04	1.895E-04	1.707E-04	1.936E-04	1.962E-04	1.773E-04	1.841E-04	1.875E-04	1.875E-04	1.739E-04	1.772E-04	1.521E-04	1.805E-04	1.838E-04	1.587E-04	1.737E-04	1.771E-04	1.520E-04	1.803E-04
2.069E-02	5.361E-02	2.072E-02	3.585E-02	4.725E-02	2.258E-02	3.648E-02	4.788E-02	2.321E-02	3.833E-02	5.353E-02	5.353E-02	3.766E-02	5.286E-02	1.996E-02	3.829E-02	5.348E-02	2.059E-02	3.758E-02	5.277E-02	1.988E-02	3.821E-02
3.593E-13	3.824E-13	3.593E-13	3.604E-13	3.776E-13	3.602E-13	3.611E-13	3.783E-13	3.610E-13	3.594E-13	3.823E-13	3.823E-13	3.586E-13	3.815E-13	3.584E-13	3.594E-13	3.823E-13	3.591E-13	3.585E-13	3.814E-13	3.583E-13	3.593E-13
3.593E-13	3.824E-13	3.593E-13	3.604E-13	3.776E-13	3.602E-13	3.611E-13	3.783E-13	3.610E-13	3.594E-13	3.823E-13	3.592E-13	3.586E-13	3.815E-13	3.584E-13	3.594E-13	3.823E-13	3.591E-13	3.585E-13	3.814E-13	3.583E-13	3.593E-13
3.595E-13	3.830E-13	3.595E-13	3.606E-13	3.781E-13	3.604E-13	3.614E-13	3.788E-13	3.612E-13	3.596E-13	3.829E-13	3.829E-13	3.588E-13	3.821E-13	3.586E-13	3.596E-13	3.828E-13	3.593E-13	3.587E-13	3.820E-13	3.585E-13	3.595E-13
1.532E-08	2.868E-08	1.548E-08	2.443E-08	2.621E-08	1.631E-08	2.485E-08	2.662E-08	1.672E-08	2.622E-08	2.859E-08	2.859E-08	2.568E-08	2.805E-08	1.485E-08	2.609E-08	2.846E-08	1.526E-08	2.558E-08	2.795E-08	1.475E-08	2.600E-08
3.812E-13	4.043E-13	3.812E-13	3.823E-13	3.995E-13	3.822E-13	3.831E-13	4.003E-13	3.829E-13	3.814E-13	4.043E-13	4.043E-13	3.806E-13	4.034E-13	3.803E-13	3.813E-13	4.042E-13	3.811E-13	3.805E-13	4.034E-13	3.802E-13	3.812E-13
1.523E-04	1.829E-04	1.541E-04	1.846E-04	1.868E-04	1.653E-04	1.893E-04	1.916E-04	1.700E-04	1.791E-04	1.821E-04	1.821E-04	1.724E-04	1.754E-04	1.467E-04	1.772E-04	1.802E-04	1.514E-04	1.717E-04	1.747E-04	1.460E-04	1.765E-04
8.550E-04	1.009E-03	8.735E-04	9.932E-04	9.965E-04	8.950E-04	1.033E-03	1.037E-03	9.352E-04	1.004E-03	1.008E-03	1.008E-03	9.371E-04	9.416E-04	8.063E-04	9.773E-04	9.819E-04	8.465E-04	9.363E-04	9.408E-04	8.055E-04	9.765E-04
1.889E-04	2.248E-04	1.934E-04	2.307E-04	2.299E-04	2.064E-04	2.372E-04	2.365E-04	2.130E-04	2.256E-04	2.246E-04	2.246E-04	2.154E-04	2.144E-04	1.830E-04	2.220E-04	2.209E-04	1.896E-04	2.152E-04	2.142E-04	1.828E-04	2.218E-04
1.163E-04	1.370E-04	1.177E-04	1.423E-04	1.416E-04	1.271E-04	1.459E-04	1.451E-04	1.306E-04	1.379E-04	1.369E-04	1.369E-04	1.333E-04	1.310E-04	1.122E-04	1.361E-04	1.351E-04	1.157E-04	1.324E-04	1.314E-04	1.120E-04	1.360E-04
1.731E-04	2.056E-04	1.755E-04	2.068E-04	2.086E-04	1.861E-04	2.128E-04	2.146E-04	1.921E-04	2.029E-04	2.054E-04	2.054E-04	1.935E-04	1.960E-04	1.659E-04	1.995E-04	2.020E-04	1.719E-04	1.933E-04	1.958E-04	1.657E-04	1.993E-04
1.383E-04	1.498E-04	1.384E-04	1.627E-04	1.585E-04	1.499E-04	1.690E-04	1.648E-04	1.562E-04	1.553E-04	1.498E-04	1.498E-04	1.481E-04	1.425E-04	1.311E-04	1.544E-04	1.488E-04	1.374E-04	1.480E-04	1.425E-04	1.310E-04	1.544E-04
2.189E-02	2.768E-02	2.190E-02	2.588E-02	2.912E-02	2.479E-02	2.628E-02	2.952E-02	2.518E-02	2.335E-02	2.767E-02	2.767E-02	2.289E-02	2.721E-02	2.143E-02	2.329E-02	2.761E-02	2.183E-02	2.288E-02	2.720E-02	2.142E-02	2.327E-02

Table 4.13 Life-cycle greenhouse gas emissions of 1 kg 65 MPa concrete

Impact category	Life-cycle assessment method Concrete mixture option Points to achieve	Unit	0-0-0 (1)0-0	0-1-0 (2)0-0.5W 0.5	0-0-1 (3)0-0.5Cs 0.5	0-0-2 (4)0-0.5S 0.5	1-0-0 (5)1F-0 1	2-0-0 (7)1S-0 1	3-0-0 (6)1B-0 1	1-1-0 (8)1F-0.5W 1.5	2-1-0 (10)1S-0.5W 1.5	3-1-0 (9)1B-0.5W 1.5	1-0-1 (11)1F-0.5Cs 1.5	2-0-1 (13)1S-0.5Cs 1.5	3-0-1 (12)1B-0.5Cs 1.5	1-0-2 (14)1F-0.5S 1.5	2-0-2 (16)1S-0.5S 1.5	3-0-2 (15)1B-0.5S 1.5
Global Warming Potential (GWP 100 years)	CML2001 - Jan. 2016, Global Warming Potential (GWP 100 years)	kg CO2 eq.	0.1810	0.1811	0.1795	0.1047	0.1431	0.1474	0.1301	0.1429	0.1472	0.1300	0.1413	0.1456	0.1284	0.1425	0.1468	0.1295
	I02+ v2.1 - Global warming 500yr - Midpoint	kg CO2 eq.	0.1807	0.1806	0.1789	0.1043	0.1418	0.1458	0.1296	0.1417	0.1458	0.1295	0.1401	0.1441	0.1281	0.1412	0.1452	0.1292
	ILCD - Climate change midpoint, incl biogenic carbon (v1.09)	kg CO2 eq.	0.1809	0.1809	0.1792	0.1047	0.1427	0.1467	0.1298	0.1426	0.1466	0.1298	0.1409	0.1449	0.1281	0.1421	0.1461	0.1292
	ReCiPe 2016 Midpoint (H) - Climate change, incl biogenic carbon	kg CO2 eq.	0.1794	0.1793	0.1777	0.1036	0.1275	0.1448	0.1286	0.1407	0.1447	0.1286	0.1390	0.1431	0.1269	0.1401	0.1442	0.1280
	TRACI 2.1, Global Warming Air, incl. biogenic carbon	kg CO2 eq.	0.1809	0.1808	0.1791	0.1046	0.1287	0.1466	0.1298	0.1426	0.1465	0.1297	0.1409	0.1448	0.1280	0.1420	0.1460	0.1292
Aquatic Ecotoxicity	CML2001 - Jan. 2016, Freshwater Aquatic Ecotoxicity Pot. (FAETP inf.)	kg DCB eq.	1.414E-04	1.410E-04	1.367E-04	8.336E-05	1.172E-04	1.242E-04	1.047E-04	1.167E-04	1.237E-04	1.043E-04	1.125E-04	1.194E-04	9.997E-05	1.160E-04	1.236E-04	1.041E-04
	I02+ v2.1 - Aquatic ecotoxicity - Midpoint	kg TEG eq. to water	3.482E-01	3.462E-01	3.254E-01	2.155E-01	3.607E-01	4.132E-01	2.737E-01	3.587E-01	4.112E-01	2.717E-01	3.379E-01	3.905E-01	2.509E-01	3.582E-01	4.109E-01	2.712E-01
	ILCD Ecotoxicity freshwater midpoint (v1.09)	CTUe	2.951E-03	2.932E-03	2.778E-03	1.803E-03	2.739E-03	2.392E-03	2.279E-03	2.720E-03	2.373E-03	2.260E-03	2.566E-03	2.219E-03	2.106E-03	2.717E-03	2.373E-03	2.258E-03
	ReCiPe 2016 Midpoint (E) - Freshwater ecotoxicity	kg 1,4 DB eq.	5.649E-06	5.614E-06	5.428E-06	3.380E-06	4.210E-06	5.198E-06	4.244E-06	4.971E-06	5.163E-06	4.209E-06	4.785E-06	4.977E-06	4.022E-06	4.978E-06	5.172E-06	4.215E-06
	TRACI 2.1, Ecotoxicity	CTUe	1.689E-03	1.679E-03	1.602E-03	9.931E-04	1.278E-03	1.269E-03	1.288E-03	1.639E-03	1.259E-03	1.278E-03	1.561E-03	1.181E-03	1.200E-03	1.639E-03	1.260E-03	1.279E-03
Eutrophication Potential (EP),	CML2001 - Jan. 2016, Eutrophication Potential (EP)	kg PO4 eq.	3.286E-05	3.297E-05	3.136E-05	1.979E-05	2.936E-05	3.045E-05	2.617E-05	2.919E-05	3.027E-05	2.599E-05	2.758E-05	2.866E-05	2.439E-05	2.873E-05	2.982E-05	2.554E-05
	I02+ v2.1 - Aquatic eutrophication - Midpoint	kg PO4 eq.	3.017E-07	2.939E-07	2.832E-07	1.672E-07	2.404E-07	4.217E-07	2.363E-07	2.326E-07	4.139E-07	2.285E-07	2.219E-07	4.032E-07	2.178E-07	2.382E-07	4.195E-07	2.341E-07
	ILCD Eutrophication freshwater midpoint (v1.09)	kg PO4 eq.	9.851E-08	9.669E-08	9.316E-08	5.499E-08	7.902E-08	1.387E-07	7.768E-08	7.648E-08	1.362E-07	7.514E-08	7.295E-08	1.327E-07	7.162E-08	7.830E-08	1.380E-07	7.696E-08
	ReCiPe 2016 Midpoint (H) - Freshwater Eutrophication	kg PO4 eq.	9.914E-08	9.660E-08	9.308E-08	5.493E-08	7.755E-08	1.386E-07	7.762E-08	7.640E-08	1.361E-07	7.507E-08	7.288E-08	1.325E-07	7.155E-08	7.822E-08	1.379E-07	7.690E-08
	TRACI 2.1, Eutrophication	kg N eq.	1.621E-05	1.616E-05	1.548E-05	1.127E-05	1.263E-05	1.669E-05	1.303E-05	1.525E-05	1.664E-05	1.299E-05	1.457E-05	1.596E-05	1.230E-05	1.507E-05	1.646E-05	1.280E-05
Human toxicity	CML2001 - Jan. 2016, Human Toxicity Potential (HTP inf.)	kg DCB eq.	3.963E-02	3.963E-02	3.957E-02	2.245E-02	2.825E-02	2.862E-02	2.782E-02	2.825E-02	2.862E-02	2.781E-02	2.819E-02	2.856E-02	2.775E-02	2.824E-02	2.861E-02	2.781E-02
	ILCD Human toxicity midpoint, cancer effects (v1.09)	CTUh	1.270E-10	1.260E-10	1.172E-10	8.118E-11	1.122E-10	1.111E-10	1.010E-10	1.112E-10	1.101E-10	1.000E-10	1.024E-10	1.013E-10	9.122E-11	1.109E-10	1.099E-10	9.969E-11
	ReCiPe 2016 Midpoint (E) - Human toxicity, cancer	kg 1,4-DB eq.	3.697E-04	3.605E-04	3.374E-04	2.325E-04	2.982E-04	3.559E-04	3.033E-04	3.337E-04	3.467E-04	2.941E-04	3.106E-04	3.236E-04	2.710E-04	3.384E-04	3.518E-04	2.968E-04
	TRACI 2.1, Human toxicity, cancer	CTUh	5.786E-11	5.769E-11	5.596E-11	3.275E-11	4.249E-11	4.682E-11	4.253E-11	4.509E-11	4.665E-11	4.236E-11	4.396E-11	4.492E-11	4.063E-11	4.501E-11	4.657E-11	4.228E-11
Photochem. Ozone Creation Potential (POCP)	CML2001 - Jan. 2016, Photochem. Ozone Creation Potential (POCP)	kg C2H4 eq. to air	2.023E-05	2.021E-05	2.192E-05	1.155E-05	1.182E-05	1.378E-05	1.130E-05	1.180E-05	1.376E-05	1.128E-05	1.351E-05	1.547E-05	1.299E-05	1.273E-05	1.469E-05	1.221E-05
	I02+ v2.1 - Photochemical oxidation - Midpoint	kg C2H4 eq. to air	9.451E-06	9.437E-06	9.087E-06	5.610E-06	7.897E-06	8.126E-06	7.075E-06	7.882E-06	8.112E-06	7.061E-06	7.532E-06	7.762E-06	6.711E-06	7.835E-06	8.067E-06	7.014E-06
	ILCD - Photochemical ozone formation midpoint, human health (v1.09)	kg NMVOC eq.	2.668E-04	2.667E-04	2.957E-04	1.535E-04	1.960E-04	2.241E-04	2.025E-04	2.210E-04	2.239E-04	2.023E-04	2.100E-04	2.129E-04	1.913E-04	2.175E-04	2.204E-04	1.988E-04
	ReCiPe 2016 Midpoint (H) - Photochemical Ozone Formation, Ecosystems	kg NOx eq.	3.539E-02	3.530E-02	3.463E-02	2.162E-02	2.608E-02	5.438E-02	2.611E-02	4.124E-02	5.430E-02	2.602E-02	4.057E-02	5.363E-02	2.585E-02	4.122E-02	5.428E-02	2.600E-02
Ozone Layer Depletion Potential	CML2001 - Jan. 2016 Ozone Layer Depletion Potential (ODP, steady state)	kg R-11 eq.	3.697E-13	3.696E-13	3.688E-13	5.242E-13	3.633E-13	3.829E-13	3.631E-13	3.632E-13	3.829E-13	3.630E-13	3.624E-13	3.821E-13	3.622E-13	3.631E-13	3.828E-13	3.629E-13
	I02+ v2.1 - Ozone layer depletion - Midpoint	kg R-11 eq.	3.697E-13	3.696E-13	3.688E-13	5.242E-13	3.633E-13	3.829E-13	3.631E-13	3.632E-13	3.829E-13	3.630E-13	3.624E-13	3.821E-13	3.622E-13	3.631E-13	3.828E-13	3.629E-13
	ILCD - Ozone depletion midpoint (v1.09)	kg CFC-11 eq.	3.700E-13	3.700E-13	3.691E-13	5.244E-13	3.635E-13	3.835E-13	3.639E-13	3.635E-13	3.835E-13	3.632E-13	3.626E-13	3.826E-13	3.624E-13	3.634E-13	3.834E-13	3.632E-13
	ReCiPe 2016 Midpoint (H) - Stratospheric Ozone Depletion	kg CFC-11 eq.	2.401E-08	2.402E-08	2.332E-08	1.628E-08	1.836E-08	3.012E-08	1.877E-08	2.799E-08	3.002E-08	1.867E-08	2.739E-08	2.943E-08	1.808E-08	2.786E-08	2.990E-08	1.855E-08
	TRACI 2.1, Ozone Depletion Air	kg CFC-11 eq.	3.916E-13	3.915E-13	3.907E-13	5.567E-13	3.850E-13	4.049E-13	3.850E-13	4.509E-11	4.048E-13	3.849E-13	3.843E-13	4.040E-13	3.841E-13	3.851E-13	4.048E-13	3.849E-13
Terrestrial Acidification	CML2001 - Jan. 2016, Acidification Potential (AP)	kg SO2 eq.	2.609E-04	2.602E-04	2.529E-04	1.508E-04	2.167E-04	2.193E-04	1.946E-04	2.160E-04	2.186E-04	1.939E-04	2.087E-04	2.113E-04	1.866E-04	2.142E-04	2.167E-04	1.920E-04
	I02+ v2.1 - Terrestrial acidification/nutrification - Midpoint	kg SO2 eq. to air	1.378E-03	1.378E-03	1.305E-03	7.989E-04	1.184E-03	1.188E-03	1.071E-03	1.183E-03	1.187E-03	1.071E-03	1.111E-03	1.114E-03	9.981E-04	1.158E-03	1.162E-03	1.045E-03
	ILCD Acidification midpoint (v1.09)	Mole of H+ eq.	3.241E-04	3.239E-04	3.129E-04	1.874E-04	2.711E-04	2.702E-04	2.433E-04	2.709E-04	2.700E-04	2.431E-04	2.599E-04	2.590E-04	2.321E-04	2.674E-04	2.665E-04	2.396E-04
	ReCiPe 2016 Midpoint (E) - Terrestrial Acidification	kg SO2 eq.	2.006E-04	2.005E-04	1.947E-04	1.161E-04	1.457E-04	1.657E-04	1.491E-04	1.665E-04	1.606E-04	1.490E-04	1.607E-04	1.598E-04	1.452E-04	1.647E-04	1.639E-04	1.472E-04
	TRACI 2.1, Acidification	kg SO2 eq.	2.885E-04	2.883E-04	2.781E-04	1.716E-04	2.123E-04	2.446E-04	2.187E-04	2.423E-04	2.444E-04	2.186E-04	2.321E-04	2.342E-04	2.083E-04	2.390E-04	2.411E-04	2.152E-04
Terrestrial ecotoxicity	CML2001 - Jan. 2016, Terrestric Ecotoxicity Potential (TETP inf.)	kg DCB eq.	2.411E-04	2.410E-04	2.338E-04	1.356E-04	1.907E-04	1.859E-04	1.760E-04	1.906E-04	1.858E-04	1.759E-04	1.834E-04	1.786E-04	1.687E-04	1.898E-04	1.850E-04	1.752E-04
	ReCiPe 2016 Midpoint (H) - Terrestrial ecotoxicity	kg 1,4-DB eq.	4.026E-02	4.025E-02	3.979E-02	2.283E-02	2.867E-02	3.364E-02	2.867E-02	2.991E-02	3.363E-02	2.866E-02	2.946E-02	3.317E-02	2.820E-02	2.986E-02	3.358E-02	2.861E-02

Life-cycle assessment method	4-0-0 (17)2F-0 2	5-0-0 (19)2S-0 2	6-0-0 (18)2B-0 2	1-1-1 (29)1F-1Cs 2	2-1-1 (31)1S-1Cs 2	3-1-1 (30)1B-1Cs 2	1-1-2 (32)1F-1S 2	2-1-2 (34)1S-1S 2	3-1-2 (33)1B-1S 2	4-1-0 (20)2F-0.5W 2.5	5-1-0 (22)2S-0.5W 2.5
CML2001 - Jan. 2016, Global Warming Potential (GWP 100 years)	0.1127	0.1361	0.1131	0.1411	0.1454	0.1282	0.1423	0.1466	0.1294	0.1302	0.1359
I02+ v2.1 - Global warming 500yr - Midpoint	0.1121	0.1342	0.1128	0.1400	0.1441	0.1278	0.1411	0.1452	0.1289	0.1288	0.1342
ILCD - Climate change midpoint, incl biogenic carbon (v1.09)	0.1123	0.1353	0.1128	0.1409	0.1449	0.1280	0.1420	0.1460	0.1292	0.1299	0.1352
ReCiPe 2016 Midpoint (H) - Climate change, incl biogenic carbon	0.1112	0.1332	0.1117	0.1390	0.1430	0.1269	0.1401	0.1441	0.1280	0.1278	0.1332
TRACI 2.1, Global Warming Air, incl. biogenic carbon	0.1123	0.1352	0.1127	0.1408	0.1448	0.1280	0.1420	0.1459	0.1291	0.1298	0.1351
CML2001 - Jan. 2016, Freshwater Aquatic Ecotoxicity Pot. (FAETP inf.)	9.151E-05	1.184E-04	9.246E-05	1.120E-04	1.190E-04	9.953E-05	1.161E-04	1.232E-04	1.036E-04	1.087E-04	1.180E-04
I02+ v2.1 - Aquatic ecotoxicity - Midpoint	2.461E-01	4.350E-01	2.488E-01	3.359E-01	3.884E-01	2.489E-01	3.562E-01	4.089E-01	2.692E-01	3.629E-01	4.330E-01
ILCD Ecotoxicity freshwater midpoint (v1.09)	2.014E-03	2.207E-03	2.055E-03	2.547E-03	2.200E-03	2.087E-03	2.698E-03	2.354E-03	2.239E-03	2.649E-03	2.188E-03
ReCiPe 2016 Midpoint (E) - Freshwater ecotoxicity	3.730E-06	5.050E-06	3.776E-06	4.750E-06	4.941E-06	3.987E-06	4.942E-06	5.137E-06	4.180E-06	4.757E-06	5.014E-06
TRACI 2.1, Ecotoxicity	1.141E-03	1.129E-03	1.154E-03	1.551E-03	1.171E-03	1.190E-03	1.629E-03	1.250E-03	1.269E-03	1.625E-03	1.119E-03
CML2001 - Jan. 2016, Eutrophication Potential (EP)	2.334E-05	2.955E-05	2.384E-05	2.740E-05	2.848E-05	2.421E-05	2.856E-05	2.964E-05	2.536E-05	2.792E-05	2.937E-05
I02+ v2.1 - Aquatic eutrophication - Midpoint	2.142E-07	4.617E-07	2.145E-07	2.141E-07	3.954E-07	2.100E-07	2.304E-07	4.117E-07	2.263E-07	2.121E-07	4.539E-07
ILCD Eutrophication freshwater midpoint (v1.09)	7.042E-08	1.519E-07	7.050E-08	7.041E-08	1.301E-07	6.907E-08	7.576E-08	1.355E-07	7.442E-08	6.974E-08	1.493E-07
ReCiPe 2016 Midpoint (H) - Freshwater Eutrophication	7.035E-08	1.517E-07	7.044E-08	7.033E-08	1.300E-07	6.901E-08	7.568E-08	1.353E-07	7.435E-08	6.966E-08	1.492E-07
TRACI 2.1, Eutrophication	1.180E-05	1.685E-05	1.197E-05	1.453E-05	1.592E-05	1.226E-05	1.503E-05	1.642E-05	1.276E-05	1.495E-05	1.680E-05
CML2001 - Jan. 2016, Human Toxicity Potential (HTP inf.)	2.387E-02	2.495E-02	2.388E-02	2.819E-02	2.856E-02	2.775E-02	2.824E-02	2.861E-02	2.780E-02	2.446E-02	2.495E-02
ILCD Human toxicity midpoint, cancer effects (v1.09)	8.968E-11	1.060E-10	9.240E-11	1.014E-10	1.003E-10	9.020E-11	1.099E-10	1.089E-10	9.867E-11	1.063E-10	1.049E-10
ReCiPe 2016 Midpoint (E) - Human toxicity, cancer	2.744E-04	3.516E-04	2.812E-04	3.013E-04	3.144E-04	2.617E-04	3.292E-04	3.426E-04	2.896E-04	3.247E-04	3.424E-04
TRACI 2.1, Human toxicity, cancer	3.737E-11	4.314E-11	3.743E-11	4.319E-11	4.475E-11	4.046E-11	4.484E-11	4.640E-11	4.211E-11	4.089E-11	4.297E-11
CML2001 - Jan. 2016, Photochem. Ozone Creation Potential (POCP)	9.125E-06	1.163E-05	8.326E-06	1.349E-05	1.545E-05	1.297E-05	1.271E-05	1.467E-05	1.219E-05	8.998E-06	1.161E-05
I02+ v2.1 - Photochemical oxidation - Midpoint	6.235E-06	7.686E-06	6.284E-06	7.518E-06	7.748E-06	6.697E-06	7.821E-06	8.053E-06	7.000E-06	7.364E-06	7.672E-06
ILCD - Photochemical ozone formation midpoint, human health (v1.09)	1.782E-04	2.098E-04	1.810E-04	2.098E-04	2.127E-04	1.912E-04	2.174E-04	2.203E-04	1.987E-04	2.058E-04	2.097E-04
ReCiPe 2016 Midpoint (H) - Photochemical Ozone Formation, Ecosystems	2.298E-02	6.071E-02	2.301E-02	4.049E-02	5.355E-02	2.527E-02	4.113E-02	5.419E-02	2.592E-02	4.322E-02	6.063E-02
CML2001 - Jan. 2016 Ozone Layer Depletion Potential (ODP, steady state)	3.608E-13	3.874E-13	3.608E-13	3.623E-13	3.820E-13	3.621E-13	3.631E-13	3.828E-13	3.629E-13	3.611E-13	3.873E-13
I02+ v2.1 - Ozone layer depletion - Midpoint	3.608E-13	3.874E-13	3.608E-13	3.623E-13	3.820E-13	3.621E-13	3.631E-13	3.828E-13	3.629E-13	3.611E-13	3.873E-13
ILCD - Ozone depletion midpoint (v1.09)	3.611E-13	3.880E-13	3.611E-13	3.626E-13	3.826E-13	3.623E-13	3.633E-13	3.833E-13	3.631E-13	3.613E-13	3.880E-13
ReCiPe 2016 Midpoint (H) - Stratospheric Ozone Depletion	1.680E-08	3.212E-08	1.698E-08	2.730E-08	2.933E-08	1.799E-08	2.777E-08	2.980E-08	1.846E-08	2.931E-08	3.202E-08
TRACI 2.1, Ozone Depletion Air	3.828E-13	4.093E-13	3.828E-13	3.842E-13	4.039E-13	3.840E-13	3.850E-13	4.047E-13	3.848E-13	3.830E-13	4.092E-13
CML2001 - Jan. 2016, Acidification Potential (AP)	1.704E-04	2.054E-04	1.725E-04	2.080E-04	2.106E-04	1.859E-04	2.134E-04	2.160E-04	1.913E-04	2.013E-04	2.047E-04
I02+ v2.1 - Terrestrial acidification/nutrification - Midpoint	9.481E-04	1.124E-03	9.692E-04	1.110E-03	1.114E-03	9.974E-04	1.157E-03	1.161E-03	1.044E-03	1.118E-03	1.124E-03
ILCD Acidification midpoint (v1.09)	2.111E-04	2.522E-04	2.163E-04	2.597E-04	2.588E-04	2.319E-04	2.672E-04	2.663E-04	2.394E-04	2.532E-04	2.520E-04
ReCiPe 2016 Midpoint (E) - Terrestrial Acidification	1.304E-04	1.541E-04	1.319E-04	1.606E-04	1.597E-04	1.431E-04	1.646E-04	1.637E-04	1.471E-04	1.551E-04	1.540E-04
TRACI 2.1, Acidification	1.927E-04	2.299E-04	1.955E-04	2.319E-04	2.340E-04	2.082E-04	2.388E-04	2.409E-04	2.151E-04	2.269E-04	2.298E-04
CML2001 - Jan. 2016, Terrestric Ecotoxicity Potential (TETP inf.)	1.543E-04	1.674E-04	1.543E-04	1.833E-04	1.785E-04	1.687E-04	1.898E-04	1.850E-04	1.751E-04	1.738E-04	1.674E-04
ReCiPe 2016 Midpoint (H) - Terrestrial ecotoxicity	2.480E-02	3.144E-02	2.481E-02	2.944E-02	3.316E-02	2.819E-02	2.985E-02	3.357E-02	2.860E-02	2.647E-02	3.142E-02

Life-cycle assessment method	6-1-0 (21)2B-0.5W 2.5	4-0-1 (23)2F-0.5Cs 2.5	5-0-1 (25)2S-0.5Cs 2.5	6-0-1 (24)2B-0.5Cs 2.5	4-0-2 (26)2F-0.5S 2.5	5-0-2 (28)2S-0.5S 2.5	6-0-2 (27)2B-0.5S 2.5	4-1-1 (35)2F-1Cs 3	5-1-1 (37)2S-1Cs 3	6-1-1 (36)2B-1Cs 3	4-1-2 (38)2F-1S 3
CML2001 - Jan. 2016, Global Warming Potential (GWP 100 years)	0.1359	0.1286	0.1343	0.1113	0.1297	0.1355	0.1125	0.1284	0.1341	0.1112	0.1295
I02+ v2.1 - Global warming 500yr - Midpoint	0.1342	0.1271	0.1325	0.1111	0.1282	0.1336	0.1122	0.1271	0.1325	0.1108	0.1282
ILCD - Climate change midpoint, incl biogenic carbon (v1.09)	0.1352	0.1282	0.1335	0.1110	0.1293	0.1347	0.1122	0.1281	0.1334	0.1110	0.1293
ReCiPe 2016 Midpoint (H) - Climate change, incl biogenic carbon	0.1332	0.1261	0.1315	0.1100	0.1273	0.1326	0.1111	0.1261	0.1315	0.1100	0.1272
TRACI 2.1, Global Warming Air, incl. biogenic carbon	0.1351	0.1281	0.1334	0.1110	0.1293	0.1346	0.1121	0.1281	0.1334	0.1109	0.1292
CML2001 - Jan. 2016, Freshwater Aquatic Ecotoxicity Pot. (FAETP inf.)	1.180E-04	1.044E-04	1.137E-04	8.774E-05	1.085E-04	1.179E-04	9.185E-05	1.039E-04	1.133E-04	8.730E-05	1.081E-04
I02+ v2.1 - Aquatic ecotoxicity - Midpoint	4.330E-01	3.421E-01	4.122E-01	2.260E-01	3.624E-01	4.328E-01	2.463E-01	3.401E-01	4.102E-01	2.240E-01	3.604E-01
ILCD Ecotoxicity freshwater midpoint (v1.09)	2.188E-03	2.495E-03	2.034E-03	1.882E-03	2.647E-03	2.190E-03	2.034E-03	2.476E-03	2.015E-03	1.863E-03	2.628E-03
ReCiPe 2016 Midpoint (E) - Freshwater ecotoxicity	5.014E-06	4.571E-06	4.826E-06	3.554E-06	4.764E-06	5.026E-06	3.747E-06	4.535E-06	4.793E-06	3.519E-06	4.728E-06
TRACI 2.1, Ecotoxicity	1.119E-03	1.547E-03	1.042E-03	1.067E-03	1.626E-03	1.121E-03	1.145E-03	1.537E-03	1.032E-03	1.057E-03	1.616E-03
CML2001 - Jan. 2016, Eutrophication Potential (EP)	2.937E-05	2.632E-05	2.776E-05	2.206E-05	2.747E-05	2.892E-05	2.322E-05	2.614E-05	2.758E-05	2.188E-05	2.729E-05
I02+ v2.1 - Aquatic eutrophication - Midpoint	4.539E-07	2.014E-07	4.432E-07	1.960E-07	2.177E-07	4.595E-07	2.123E-07	1.936E-07	4.354E-07	1.882E-07	2.099E-07
ILCD Eutrophication freshwater midpoint (v1.09)	1.493E-07	6.622E-08	1.458E-07	6.443E-08	7.157E-08	1.512E-07	6.978E-08	6.367E-08	1.433E-07	6.189E-08	6.902E-08
ReCiPe 2016 Midpoint (H) - Freshwater Eutrophication	1.492E-07	6.614E-08	1.457E-07	6.438E-08	7.149E-08	1.510E-07	6.972E-08	6.360E-08	1.431E-07	6.183E-08	6.894E-08
TRACI 2.1, Eutrophication	1.680E-05	1.427E-05	1.612E-05	1.125E-05	1.476E-05	1.662E-05	1.175E-05	1.422E-05	1.608E-05	1.120E-05	1.472E-05
CML2001 - Jan. 2016, Human Toxicity Potential (HTP inf.)	2.495E-02	2.440E-02	2.489E-02	2.382E-02	2.445E-02	2.494E-02	2.387E-02	2.440E-02	2.489E-02	2.381E-02	2.445E-02
ILCD Human toxicity midpoint, cancer effects (v1.09)	1.049E-10	9.748E-11	9.613E-11	8.257E-11	1.060E-10	1.049E-10	9.104E-11	9.646E-11	9.510E-11	8.155E-11	1.049E-10
ReCiPe 2016 Midpoint (E) - Human toxicity, cancer	3.424E-04	3.016E-04	3.193E-04	2.488E-04	3.295E-04	3.478E-04	2.767E-04	2.924E-04	3.100E-04	2.396E-04	3.202E-04
TRACI 2.1, Human toxicity, cancer	4.297E-11	3.916E-11	4.124E-11	3.552E-11	4.081E-11	4.290E-11	3.717E-11	3.899E-11	4.107E-11	3.535E-11	4.064E-11
CML2001 - Jan. 2016, Photochem. Ozone Creation Potential (POCP)	1.161E-05	1.070E-05	1.332E-05	1.001E-05	9.923E-06	1.254E-05	8.725E-06	1.069E-05	1.330E-05	9.997E-06	9.906E-06
I02+ v2.1 - Photochemical oxidation - Midpoint	7.672E-06	7.014E-06	7.322E-06	5.919E-06	7.317E-06	7.628E-06	6.222E-06	7.000E-06	7.308E-06	5.905E-06	7.303E-06
ILCD - Photochemical ozone formation midpoint, human health (v1.09)	2.097E-04	1.948E-04	1.986E-04	1.698E-04	2.023E-04	2.062E-04	1.774E-04	1.946E-04	1.985E-04	1.697E-04	2.022E-04
ReCiPe 2016 Midpoint (H) - Photochemical Ozone Formation, Ecosystems	6.063E-02	4.255E-02	5.997E-02	2.226E-02	4.319E-02	6.061E-02	2.291E-02	4.247E-02	5.988E-02	2.218E-02	4.311E-02
CML2001 - Jan. 2016 Ozone Layer Depletion Potential (ODP, steady state)	3.873E-13	3.602E-13	3.865E-13	3.600E-13	3.610E-13	3.872E-13	3.607E-13	3.602E-13	3.864E-13	3.599E-13	3.609E-13
I02+ v2.1 - Ozone layer depletion - Midpoint	3.608E-13	3.602E-13	3.865E-13	3.600E-13	3.610E-13	3.872E-13	3.607E-13	3.602E-13	3.864E-13	3.599E-13	3.609E-13
ILCD - Ozone depletion midpoint (v1.09)	3.880E-13	3.605E-13	3.871E-13	3.602E-13	3.612E-13	3.879E-13	3.610E-13	3.604E-13	3.871E-13	3.601E-13	3.612E-13
ReCiPe 2016 Midpoint (H) - Stratospheric Ozone Depletion	3.202E-08	2.872E-08	3.143E-08	1.630E-08	2.919E-08	3.190E-08	1.677E-08	2.862E-08	3.133E-08	1.620E-08	2.909E-08
TRACI 2.1, Ozone Depletion Air	4.092E-13	3.822E-13	4.084E-13	3.819E-13	3.829E-13	4.092E-13	3.827E-13	3.821E-13	4.084E-13	3.818E-13	3.829E-13
CML2001 - Jan. 2016, Acidification Potential (AP)	2.047E-04	1.940E-04	1.975E-04	1.645E-04	1.994E-04	2.029E-04	1.699E-04	1.933E-04	1.968E-04	1.638E-04	1.987E-04
I02+ v2.1 - Terrestrial acidification/nutrification - Midpoint	1.124E-03	1.046E-03	1.051E-03	8.959E-04	1.093E-03	1.098E-03	9.429E-04	1.045E-03	1.050E-03	8.951E-04	1.092E-03
ILCD Acidification midpoint (v1.09)	2.520E-04	2.422E-04	2.410E-04	2.051E-04	2.497E-04	2.485E-04	2.126E-04	2.420E-04	2.408E-04	2.049E-04	2.495E-04
ReCiPe 2016 Midpoint (E) - Terrestrial Acidification	1.540E-04	1.501E-04	1.476E-04	1.260E-04	1.534E-04	1.522E-04	1.300E-04	1.492E-04	1.481E-04	1.259E-04	1.532E-04
TRACI 2.1, Acidification	2.298E-04	2.167E-04	2.195E-04	1.851E-04	2.236E-04	2.265E-04	1.920E-04	2.165E-04	2.194E-04	1.849E-04	2.234E-04
CML2001 - Jan. 2016, Terrestric Ecotoxicity Potential (TETP inf.)	1.674E-04	1.666E-04	1.602E-04	1.470E-04	1.730E-04	1.666E-04	1.535E-04	1.665E-04	1.601E-04	1.470E-04	1.729E-04
ReCiPe 2016 Midpoint (H) - Terrestrial ecotoxicity	3.142E-02	2.601E-02	3.097E-02	2.434E-02	2.642E-02	3.137E-02	2.475E-02	2.600E-02	3.095E-02	2.433E-02	2.640E-02

Table 4.14 Life-cycle greenhouse gas emissions of 1 kg 80 MPa concrete

Impact category	Life-cycle assessment method Concrete mixture option Points to achieve	Unit	0-0-0 (1)0-0	0-1-0 (2)0-0.5W 0.5	0-0-1 (3)0-0.5Cs 0.5	0-0-2 (4)0-0.5S 0.5	1-0-0 (5)1F-0 1	2-0-0 (7)1S-0 1	3-0-0 (6)1B-0 1	1-1-0 (8)1F-0.5W 1.5	2-1-0 (10)1S-0.5W 1.5	3-1-0 (9)1B-0.5W 1.5	1-0-1 (11)1F-0.5Cs 1.5	2-0-1 (13)1S-0.5Cs 1.5	3-0-1 (12)1B-0.5Cs 1.5	1-0-2 (14)1F-0.5S 1.5	2-0-2 (16)1S-0.5S 1.5	3-0-2 (15)1B-0.5S 1.5
Global Warming Potential (GWP 100 years)	CML2001 - Jan. 2016, Global Warming Potential (GWP 100 years)	kg CO_2 eq.	0.2019	0.2020	0.2003	0.1049	0.1595	0.1642	0.1449	0.1592	0.1640	0.1447	0.1575	0.1623	0.1430	0.1588	0.1636	0.1443
	IO2+ v2.1 - Global warming 500yr - Midpoint	kg CO_2 eq.	0.2016	0.2015	0.1997	0.1045	0.1580	0.1626	0.1444	0.1580	0.1625	0.1443	0.1562	0.1607	0.1428	0.1574	0.1620	0.1440
	ILCD - Climate change midpoint, incl biogenic carbon (v1.09)	kg CO_2 eq.	0.2018	0.2018	0.2000	0.1048	0.1590	0.1635	0.1446	0.1589	0.1634	0.1445	0.1571	0.1616	0.1427	0.1584	0.1629	0.1440
	ReCiPe 2016 Midpoint (H) - Climate change, incl biogenic carbon	kg CO_2 eq.	0.2001	0.2001	0.1983	0.1038	0.1420	0.1614	0.1433	0.1568	0.1613	0.1432	0.1550	0.1595	0.1414	0.1563	0.1608	0.1427
	TRACI 2.1, Global Warming Air, incl. biogenic carbon	kg CO_2 eq.	0.2018	0.2017	0.1999	0.1048	0.1433	0.1634	0.1445	0.1589	0.1633	0.1445	0.1571	0.1615	0.1427	0.1583	0.1628	0.1439
Aquatic Ecotoxicity	CML2001 - Jan. 2016, Freshwater Aquatic Ecotoxicity Pot. (FAETP inf.)	kg DCB eq.	1.560E-04	1.556E-04	1.513E-04	8.346E-05	1.288E-04	1.367E-04	1.149E-04	1.284E-04	1.362E-04	1.144E-04	1.241E-04	1.320E-04	1.102E-04	1.283E-04	1.362E-04	1.144E-04
	IO2+ v2.1 - Aquatic ecotoxicity - Midpoint	kg TEG eq. to water	3.771E-01	3.751E-01	3.544E-01	2.190E-01	3.911E-01	4.500E-01	2.936E-01	3.891E-01	4.480E-01	2.916E-01	3.684E-01	4.273E-01	2.709E-01	3.891E-01	4.481E-01	2.915E-01
	ILCD Ecotoxicity freshwater midpoint (v1.09)	CTUe	3.218E-03	3.199E-03	3.045E-03	1.810E-03	2.979E-03	2.591E-03	2.464E-03	2.961E-03	2.572E-03	2.445E-03	2.807E-03	2.419E-03	2.292E-03	2.962E-03	2.576E-03	2.447E-03
	ReCiPe 2016 Midpoint (E) - Freshwater ecotoxicity	kg 1,4 DB eq.	6.207E-06	6.172E-06	5.986E-06	3.393E-06	4.593E-06	5.702E-06	4.632E-06	5.451E-06	5.666E-06	4.597E-06	5.266E-06	5.481E-06	4.411E-06	5.463E-06	5.681E-06	4.608E-06
	TRACI 2.1, Ecotoxicity	CTUe	1.848E-03	1.838E-03	1.761E-03	9.987E-04	1.387E-03	1.376E-03	1.398E-03	1.792E-03	1.366E-03	1.388E-03	1.715E-03	1.289E-03	1.311E-03	1.794E-03	1.370E-03	1.390E-03
Eutrophication Potential (EP)	CML2001 - Jan. 2016, Eutrophication Potential (EP)	kg PO_4 eq.	3.603E-05	3.614E-05	3.440E-05	2.006E-05	3.210E-05	3.331E-05	2.852E-05	3.192E-05	3.313E-05	2.834E-05	3.018E-05	3.139E-05	2.660E-05	3.149E-05	3.270E-05	2.791E-05
	IO2+ v2.1 - Aquatic eutrophication - Midpoint	kg PO_4 eq.	3.264E-07	3.187E-07	3.080E-07	1.690E-07	2.576E-07	4.609E-07	2.531E-07	2.499E-07	4.531E-07	2.453E-07	2.392E-07	4.424E-07	2.347E-07	2.558E-07	4.591E-07	2.513E-07
	ILCD Eutrophication freshwater midpoint (v1.09)	kg PO_4 eq.	1.068E-07	1.048E-07	1.013E-07	5.556E-08	8.472E-08	1.516E-07	8.322E-08	8.218E-08	1.491E-07	8.068E-08	7.867E-08	1.456E-07	7.717E-08	8.412E-08	1.510E-07	8.262E-08
	ReCiPe 2016 Midpoint (H) - Freshwater Eutrophication	kg PO_4 eq.	1.073E-07	1.047E-07	1.012E-07	5.550E-08	8.307E-08	1.515E-07	8.314E-08	8.209E-08	1.490E-07	8.061E-08	7.858E-08	1.454E-07	7.710E-08	8.403E-08	1.509E-07	8.255E-08
	TRACI 2.1, Eutrophication	kg N eq.	1.760E-05	1.756E-05	1.682E-05	1.170E-05	1.359E-05	1.814E-05	1.405E-05	1.654E-05	1.810E-05	1.401E-05	1.581E-05	1.737E-05	1.327E-05	1.637E-05	1.792E-05	1.383E-05
Human toxicity	CML2001 - Jan. 2016, Human Toxicity Potential (HTP inf.)	kg DCB eq.	4.438E-02	4.438E-02	4.432E-02	2.238E-02	3.163E-02	3.205E-02	3.114E-02	3.163E-02	3.204E-02	3.114E-02	3.157E-02	3.198E-02	3.108E-02	3.163E-02	3.204E-02	3.113E-02
	ILCD Human toxicity midpoint, cancer effects (v1.09)	CTUh	1.372E-10	1.362E-10	1.274E-10	8.144E-11	1.207E-10	1.195E-10	1.082E-10	1.197E-10	1.184E-10	1.071E-10	1.109E-10	1.097E-10	9.835E-11	1.196E-10	1.185E-10	1.070E-10
	ReCiPe 2016 Midpoint (E) - Human toxicity, cancer	kg 1,4-DB eq.	3.951E-04	3.859E-04	3.628E-04	2.349E-04	3.150E-04	3.796E-04	3.206E-04	3.558E-04	3.704E-04	3.114E-04	3.328E-04	3.474E-04	2.884E-04	3.613E-04	3.763E-04	3.169E-04
	TRACI 2.1, Human toxicity, cancer	CTUh	6.386E-11	6.368E-11	6.196E-11	3.274E-11	4.663E-11	5.148E-11	4.668E-11	4.957E-11	5.131E-11	4.651E-11	4.784E-11	4.959E-11	4.478E-11	4.953E-11	5.128E-11	4.647E-11
Photochem. Ozone Creation Potential (POCP)	CML2001 - Jan. 2016, Photochem. Ozone Creation Potential (POCP)	kg C_2H_4 eq. to air	2.307E-05	2.305E-05	2.500E-05	1.140E-05	1.360E-05	1.580E-05	1.302E-05	1.359E-05	1.578E-05	1.301E-05	1.554E-05	1.774E-05	1.496E-05	1.452E-05	1.672E-05	1.394E-05
	IO2+ v2.1 - Photochemical oxidation - Midpoint	kg C_2H_4 eq. to air	1.040E-05	1.038E-05	1.009E-05	5.638E-06	8.654E-06	8.912E-06	7.734E-06	8.640E-06	8.898E-06	7.720E-06	8.287E-06	8.544E-06	7.367E-06	8.601E-06	8.860E-06	7.680E-06
	ILCD - Photochemical ozone formation midpoint, human health (v1.09)	kg NMVOC eq.	2.946E-04	2.944E-04	2.827E-04	1.539E-04	2.152E-04	2.468E-04	2.226E-04	2.434E-04	2.466E-04	2.224E-04	2.316E-04	2.348E-04	2.106E-04	2.400E-04	2.432E-04	2.191E-04
	ReCiPe 2016 Midpoint (H) - Photochemical Ozone Formation, Ecosystems	kg NO_x eq.	3.906E-02	3.897E-02	3.830E-02	2.190E-02	2.862E-02	6.034E-02	2.866E-02	4.563E-02	6.026E-02	2.857E-02	4.496E-02	5.959E-02	2.790E-02	4.562E-02	6.026E-02	2.857E-02
Ozone Layer Depletion Potential	CML2001 - Jan. 2016 Ozone Layer Depletion Potential (ODP, steady state)	kg R-11 eq.	3.722E-13	3.722E-13	3.713E-13	5.858E-13	3.651E-13	3.871E-13	3.648E-13	3.650E-13	3.870E-13	3.648E-13	3.642E-13	3.862E-13	3.639E-13	3.650E-13	3.870E-13	3.647E-13
	IO2+ v2.1 - Ozone layer depletion - Midpoint	kg R-11 eq.	3.722E-13	3.722E-13	3.713E-13	5.858E-13	3.651E-13	3.871E-13	3.648E-13	3.650E-13	3.870E-13	3.648E-13	3.642E-13	3.862E-13	3.639E-13	3.650E-13	3.870E-13	3.647E-13
	ILCD - Ozone depletion midpoint (v1.09)	kg CFC-11 eq.	3.726E-13	3.726E-13	3.718E-13	5.860E-13	3.654E-13	3.878E-13	3.651E-13	3.653E-13	3.877E-13	3.650E-13	3.645E-13	3.869E-13	3.642E-13	3.653E-13	3.877E-13	3.650E-13
	ReCiPe 2016 Midpoint (H) - Stratospheric Ozone Depletion	kg CFC-11 eq.	2.628E-08	2.629E-08	2.555E-08	1.682E-08	1.994E-08	3.312E-08	2.040E-08	3.075E-08	3.303E-08	2.031E-08	3.011E-08	3.239E-08	1.967E-08	3.063E-08	3.291E-08	2.019E-08
	TRACI 2.1, Ozone Depletion Air	kg CFC-11 eq.	3.942E-13	3.941E-13	3.933E-13	6.222E-13	3.868E-13	4.091E-13	3.868E-13	4.957E-11	4.090E-13	3.867E-13	3.861E-13	4.082E-13	3.859E-13	3.869E-13	4.090E-13	3.867E-13
Terrestrial Acidification	CML2001 - Jan. 2016, Acidification Potential (AP)	kg SO_2 eq.	2.888E-04	2.881E-04	2.804E-04	1.508E-04	2.394E-04	2.423E-04	2.146E-04	2.387E-04	2.416E-04	2.139E-04	2.310E-04	2.338E-04	2.062E-04	2.370E-04	2.399E-04	2.122E-04
	IO2+ v2.1 - Terrestrial acidification/nutrification - Midpoint	kg SO_2 eq. to air	1.517E-03	1.516E-03	1.438E-03	7.976E-04	1.300E-03	1.305E-03	1.174E-03	1.300E-03	1.304E-03	1.173E-03	1.221E-03	1.225E-03	1.095E-03	1.275E-03	1.279E-03	1.149E-03
	ILCD Acidification midpoint (v1.09)	Mole of H^+ eq.	3.587E-04	3.585E-04	3.467E-04	1.880E-04	2.994E-04	2.983E-04	2.682E-04	2.992E-04	2.981E-04	2.680E-04	2.874E-04	2.863E-04	2.562E-04	2.958E-04	2.947E-04	2.646E-04
	ReCiPe 2016 Midpoint (E) - Terrestrial Acidification	kg SO_2 eq.	2.223E-04	2.222E-04	2.161E-04	1.165E-04	1.608E-04	1.833E-04	1.646E-04	1.841E-04	1.831E-04	1.645E-04	1.779E-04	1.770E-04	1.583E-04	1.824E-04	1.815E-04	1.628E-04
	TRACI 2.1, Acidification	kg SO_2 eq.	3.185E-04	3.183E-04	3.074E-04	1.731E-04	2.331E-04	2.694E-04	2.404E-04	2.668E-04	2.692E-04	2.402E-04	2.558E-04	2.582E-04	2.293E-04	2.636E-04	2.660E-04	2.370E-04
Terrestrial ecotoxicity	CML2001 - Jan. 2016, Terrestric Ecotoxicity Potential (TETP inf.)	kg DCB eq.	2.667E-04	2.666E-04	2.594E-04	1.355E-04	2.101E-04	2.048E-04	1.937E-04	2.101E-04	2.047E-04	1.937E-04	2.029E-04	1.975E-04	1.865E-04	2.095E-04	2.041E-04	1.930E-04
	ReCiPe 2016 Midpoint (H) - Terrestrial ecotoxicity	kg 1,4-DB eq.	4.488E-02	4.487E-02	4.442E-02	2.278E-02	3.189E-02	3.746E-02	3.190E-02	3.329E-02	3.745E-02	3.188E-02	3.283E-02	3.700E-02	3.143E-02	3.325E-02	3.741E-02	3.184E-02

4-0-0 (17)2F-0 2	5-0-0 (19)2S-0 2	6-0-0 (18)2B-0 2	1-1-1 (29)1F-1Cs 2	2-1-1 (31)1S-1Cs 2	3-1-1 (30)1B-1Cs 2	1-1-2 (32)1F-1S 2	2-1-2 (34)1S-1S 2	3-1-2 (33)1B-1S 2	4-1-0 (20)2F-0.5W 2.5	5-1-0 (22)2S-0.5W 2.5	6-1-0 (21)2B-0.5W 2.5	4-0-1 (23)2F-0.5Cs 2.5	5-0-1 (25)2S-0.5Cs 2.5	6-0-1 (24)2B-0.5Cs 2.5	4-0-2 (26)2F-0.5S 2.5	5-0-2 (28)2S-0.5S 2.5	6-0-2 (27)2B-0.5S 2.5	4-1-1 (35)2F-1Cs 3	5-1-1 (37)2S-1Cs 3	6-1-1 (36)2B-1Cs 3	4-1-2 (38)2F-1S 3
0.1254	0.1515	0.1258	0.1573	0.1621	0.1428	0.1586	0.1634	0.1441	0.1450	0.1514	0.1514	0.1432	0.1496	0.1239	0.1445	0.1510	0.1252	0.1431	0.1495	0.1237	0.1444
0.1247	0.1496	0.1256	0.1561	0.1607	0.1425	0.1574	0.1619	0.1437	0.1435	0.1495	0.1495	0.1417	0.1477	0.1237	0.1429	0.1490	0.1250	0.1416	0.1477	0.1234	0.1429
0.1249	0.1507	0.1255	0.1570	0.1615	0.1426	0.1583	0.1628	0.1439	0.1447	0.1506	0.1506	0.1428	0.1488	0.1236	0.1441	0.1501	0.1249	0.1428	0.1487	0.1236	0.1441
0.1238	0.1485	0.1243	0.1550	0.1595	0.1414	0.1562	0.1608	0.1427	0.1424	0.1484	0.1484	0.1406	0.1466	0.1225	0.1418	0.1479	0.1237	0.1405	0.1466	0.1224	0.1418
0.1249	0.1506	0.1255	0.1570	0.1615	0.1426	0.1583	0.1628	0.1439	0.1446	0.1505	0.1505	0.1428	0.1487	0.1236	0.1441	0.1500	0.1249	0.1427	0.1487	0.1235	0.1440
1.001E-04	1.303E-04	1.011E-04	1.237E-04	1.315E-04	1.097E-04	1.279E-04	1.358E-04	1.139E-04	1.194E-04	1.298E-04	1.298E-04	1.151E-04	1.256E-04	9.644E-05	1.193E-04	1.299E-04	1.006E-04	1.147E-04	1.251E-04	9.601E-05	1.189E-04
2.627E-01	4.744E-01	2.657E-01	3.664E-01	4.253E-01	2.688E-01	3.871E-01	4.461E-01	2.895E-01	3.938E-01	4.724E-01	4.724E-01	3.731E-01	4.517E-01	2.430E-01	3.937E-01	4.727E-01	2.636E-01	3.711E-01	4.497E-01	2.410E-01	3.917E-01
2.168E-03	2.384E-03	2.213E-03	2.788E-03	2.400E-03	2.273E-03	2.943E-03	2.557E-03	2.428E-03	2.881E-03	2.365E-03	2.365E-03	2.728E-03	2.212E-03	2.041E-03	2.883E-03	2.371E-03	2.196E-03	2.709E-03	2.193E-03	2.022E-03	2.864E-03
4.055E-06	5.535E-06	4.107E-06	5.231E-06	5.445E-06	4.376E-06	5.428E-06	5.645E-06	4.573E-06	5.211E-06	5.500E-06	5.500E-06	5.026E-06	5.312E-06	3.886E-06	5.223E-06	5.517E-06	4.083E-06	4.991E-06	5.279E-06	3.851E-06	5.188E-06
1.233E-03	1.220E-03	1.248E-03	1.705E-03	1.279E-03	1.301E-03	1.784E-03	1.360E-03	1.380E-03	1.777E-03	1.210E-03	1.210E-03	1.700E-03	1.133E-03	1.161E-03	1.779E-03	1.214E-03	1.240E-03	1.690E-03	1.123E-03	1.151E-03	1.769E-03
2.536E-05	3.231E-05	2.592E-05	3.000E-05	3.121E-05	2.642E-05	3.131E-05	3.252E-05	2.773E-05	3.051E-05	3.213E-05	3.213E-05	2.877E-05	3.039E-05	2.400E-05	3.008E-05	3.170E-05	2.531E-05	2.859E-05	3.021E-05	2.382E-05	2.990E-05
2.284E-07	5.057E-07	2.287E-07	2.315E-07	4.347E-07	2.269E-07	2.481E-07	4.513E-07	2.435E-07	2.270E-07	4.979E-07	4.979E-07	2.163E-07	4.873E-07	2.102E-07	2.329E-07	5.039E-07	2.268E-07	2.085E-07	4.795E-07	2.025E-07	2.251E-07
7.507E-08	1.664E-07	7.517E-08	7.613E-08	1.430E-07	7.463E-08	8.159E-08	1.485E-07	8.009E-08	7.463E-08	1.639E-07	1.639E-07	7.112E-08	1.603E-07	6.912E-08	7.657E-08	1.658E-07	7.457E-08	6.858E-08	1.578E-07	6.658E-08	7.404E-08
7.500E-08	1.662E-07	7.510E-08	7.605E-08	1.429E-07	7.456E-08	8.150E-08	1.484E-07	8.001E-08	7.455E-08	1.637E-07	1.637E-07	7.104E-08	1.602E-07	6.906E-08	7.649E-08	1.656E-07	7.451E-08	6.850E-08	1.577E-07	6.652E-08	7.395E-08
1.267E-05	1.833E-05	1.286E-05	1.577E-05	1.733E-05	1.323E-05	1.632E-05	1.788E-05	1.379E-05	1.621E-05	1.828E-05	1.828E-05	1.547E-05	1.755E-05	1.209E-05	1.603E-05	1.811E-05	1.264E-05	1.543E-05	1.751E-05	1.205E-05	1.599E-05
2.672E-02	2.793E-02	2.673E-02	3.157E-02	3.198E-02	3.108E-02	3.162E-02	3.203E-02	3.113E-02	2.738E-02	2.793E-02	2.793E-02	2.732E-02	2.787E-02	2.667E-02	2.738E-02	2.793E-02	2.672E-02	2.732E-02	2.787E-02	2.666E-02	2.737E-02
9.541E-11	1.137E-10	9.846E-11	1.099E-10	1.086E-10	9.734E-11	1.185E-10	1.175E-10	1.060E-10	1.142E-10	1.126E-10	1.126E-10	1.054E-10	1.039E-10	8.866E-11	1.141E-10	1.129E-10	9.734E-11	1.044E-10	1.028E-10	8.764E-11	1.130E-10
2.882E-04	3.748E-04	2.958E-04	3.236E-04	3.382E-04	2.792E-04	3.521E-04	3.671E-04	3.077E-04	3.458E-04	3.656E-04	3.656E-04	3.228E-04	3.425E-04	2.636E-04	3.513E-04	3.718E-04	2.921E-04	3.135E-04	3.333E-04	2.544E-04	3.421E-04
4.089E-11	4.736E-11	4.096E-11	4.767E-11	4.942E-11	4.461E-11	4.936E-11	5.111E-11	4.630E-11	4.486E-11	4.719E-11	4.719E-11	4.313E-11	4.547E-11	3.906E-11	4.482E-11	4.716E-11	4.075E-11	4.296E-11	4.530E-11	3.889E-11	4.465E-11
1.057E-05	1.338E-05	9.676E-06	1.552E-05	1.772E-05	1.494E-05	1.450E-05	1.670E-05	1.392E-05	1.043E-05	1.336E-05	1.336E-05	1.239E-05	1.532E-05	1.161E-05	1.136E-05	1.430E-05	1.008E-05	1.237E-05	1.530E-05	1.159E-05	1.134E-05
6.792E-06	8.418E-06	6.847E-06	8.273E-06	8.530E-06	7.353E-06	8.586E-06	8.846E-06	7.666E-06	8.059E-06	8.404E-06	8.404E-06	7.706E-06	8.051E-06	6.479E-06	8.020E-06	8.369E-06	6.793E-06	7.692E-06	8.037E-06	6.465E-06	8.006E-06
1.954E-04	2.308E-04	1.986E-04	2.315E-04	2.347E-04	2.105E-04	2.399E-04	2.431E-04	2.189E-04	2.264E-04	2.307E-04	2.307E-04	2.146E-04	2.189E-04	1.866E-04	2.230E-04	2.273E-04	1.951E-04	2.144E-04	2.187E-04	1.865E-04	2.228E-04
2.515E-02	6.744E-02	2.519E-02	4.488E-02	5.951E-02	2.782E-02	4.554E-02	6.018E-02	2.849E-02	4.785E-02	6.736E-02	6.796E-02	4.718E-02	6.670E-02	2.444E-02	4.784E-02	6.736E-02	2.510E-02	4.709E-02	6.661E-02	2.436E-02	4.776E-02
3.623E-13	3.921E-13	3.623E-13	3.641E-13	3.862E-13	3.639E-13	3.649E-13	3.869E-13	3.646E-13	3.626E-13	3.920E-13	3.920E-13	3.618E-13	3.912E-13	3.615E-13	3.626E-13	3.920E-13	3.622E-13	3.617E-13	3.911E-13	3.614E-13	3.625E-13
3.624E-13	3.921E-13	3.624E-13	3.641E-13	3.862E-13	3.639E-13	3.649E-13	3.869E-13	3.646E-13	3.626E-13	3.920E-13	3.623E-13	3.618E-13	3.912E-13	3.615E-13	3.626E-13	3.920E-13	3.622E-13	3.617E-13	3.911E-13	3.614E-13	3.625E-13
3.626E-13	3.928E-13	3.626E-13	3.644E-13	3.868E-13	3.642E-13	3.652E-13	3.876E-13	3.649E-13	3.629E-13	3.927E-13	3.927E-13	3.620E-13	3.919E-13	3.617E-13	3.628E-13	3.927E-13	3.625E-13	3.620E-13	3.919E-13	3.616E-13	3.628E-13
1.820E-08	3.536E-08	1.841E-08	3.001E-08	3.229E-08	1.958E-08	3.054E-08	3.282E-08	2.010E-08	3.223E-08	3.527E-08	3.527E-08	3.159E-08	3.463E-08	1.767E-08	3.211E-08	3.516E-08	1.820E-08	3.150E-08	3.454E-08	1.758E-08	3.202E-08
3.843E-13	4.140E-13	3.843E-13	3.860E-13	4.081E-13	3.858E-13	3.868E-13	4.089E-13	3.866E-13	3.845E-13	4.140E-13	4.140E-13	3.837E-13	4.131E-13	3.834E-13	3.845E-13	4.139E-13	3.842E-13	3.837E-13	4.131E-13	3.833E-13	3.844E-13
1.876E-04	2.268E-04	1.899E-04	2.302E-04	2.331E-04	2.054E-04	2.363E-04	2.392E-04	2.115E-04	2.223E-04	2.261E-04	2.261E-04	2.145E-04	2.184E-04	1.814E-04	2.205E-04	2.244E-04	1.874E-04	2.138E-04	2.176E-04	1.807E-04	2.198E-04
1.036E-03	1.234E-03	1.060E-03	1.220E-03	1.225E-03	1.094E-03	1.274E-03	1.278E-03	1.148E-03	1.227E-03	1.233E-03	1.233E-03	1.149E-03	1.155E-03	9.807E-04	1.202E-03	1.208E-03	1.034E-03	1.148E-03	1.154E-03	9.799E-04	1.202E-03
2.321E-04	2.782E-04	2.380E-04	2.872E-04	2.861E-04	2.560E-04	2.956E-04	2.945E-04	2.644E-04	2.794E-04	2.780E-04	2.780E-04	2.676E-04	2.662E-04	2.260E-04	2.760E-04	2.747E-04	2.344E-04	2.674E-04	2.660E-04	2.258E-04	2.758E-04
1.437E-04	1.702E-04	1.454E-04	1.778E-04	1.769E-04	1.582E-04	1.823E-04	1.813E-04	1.627E-04	1.714E-04	1.701E-04	1.701E-04	1.660E-04	1.632E-04	1.391E-04	1.697E-04	1.685E-04	1.436E-04	1.651E-04	1.638E-04	1.390E-04	1.696E-04
2.112E-04	2.530E-04	2.144E-04	2.557E-04	2.580E-04	2.291E-04	2.634E-04	2.658E-04	2.368E-04	2.496E-04	2.528E-04	2.528E-04	2.387E-04	2.419E-04	2.032E-04	2.464E-04	2.496E-04	2.110E-04	2.385E-04	2.417E-04	2.030E-04	2.462E-04
1.694E-04	1.841E-04	1.694E-04	2.028E-04	1.974E-04	1.864E-04	2.094E-04	2.040E-04	1.930E-04	1.912E-04	1.841E-04	1.841E-04	1.840E-04	1.769E-04	1.622E-04	1.906E-04	1.834E-04	1.687E-04	1.840E-04	1.768E-04	1.621E-04	1.905E-04
2.755E-02	3.499E-02	2.757E-02	3.282E-02	3.698E-02	3.141E-02	3.323E-02	3.740E-02	3.183E-02	2.943E-02	3.498E-02	3.498E-02	2.897E-02	3.452E-02	2.710E-02	2.939E-02	3.494E-02	2.751E-02	2.896E-02	3.451E-02	2.708E-02	2.937E-02

Table 4.15 Life-cycle greenhouse gas emissions of 1 kg 100 MPa concrete

Impact category	Life-cycle assessment method Concrete mixture option Points to achieve	Unit	0-0-0 (1)0-0	0-1-0 (2)0-0.5W 0.5	0-0-1 (3)0-0.5Cs 0.5	0-0-2 (4)0-0.5S 0.5	1-0-0 (5)1F-0 1	2-0-0 (7)1S-0 1	3-0-0 (6)1B-0 1	1-1-0 (8)1F-0.5W 1.5	2-1-0 (10)1S-0.5W 1.5	3-1-0 (9)1B-0.5W 1.5	1-0-1 (11)1F-0.5Cs 1.5	2-0-1 (13)1S-0.5Cs 1.5	3-0-1 (12)1B-0.5Cs 1.5	1-0-2 (14)1F-0.5S 1.5	2-0-2 (16)1S-0.5S 1.5	3-0-2 (15)1B-0.5S 1.5
Global Warming Potential (GWP 100 years)	CML2001 - Jan. 2016, Global Warming Potential (GWP 100 years)	kg CO2 eq.	0.2157	0.2158	0.2140	0.2156	0.1703	0.1753	0.1547	0.1700	0.1751	0.1545	0.1682	0.1733	0.1527	0.1696	0.1748	0.1541
	IO2+ v2.1 - Global warming 500yr - Midpoint	kg CO2 eq.	0.2154	0.2154	0.2135	0.2152	0.1688	0.1736	0.1541	0.1687	0.1736	0.1541	0.1668	0.1717	0.1524	0.1682	0.1730	0.1538
	ILCD - Climate change midpoint, incl biogenic carbon (v1.09)	kg CO2 eq.	0.2157	0.2156	0.2137	0.2155	0.1698	0.1746	0.1543	0.1697	0.1745	0.1543	0.1678	0.1726	0.1524	0.1692	0.1740	0.1538
	ReCiPe 2016 Midpoint (H) - Climate change, incl biogenic carbon	kg CO2 eq.	0.2139	0.2138	0.2119	0.2137	0.1516	0.1723	0.1530	0.1675	0.1723	0.1529	0.1656	0.1704	0.1510	0.1669	0.1718	0.1524
	TRACI 2.1, Global Warming Air, incl. biogenic carbon	kg CO2 eq.	0.2156	0.2155	0.2136	0.2154	0.1529	0.1745	0.1543	0.1697	0.1744	0.1542	0.1677	0.1725	0.1523	0.1691	0.1739	0.1537
Aquatic Ecotoxicity	CML2001 - Jan. 2016, Freshwater Aquatic Ecotoxicity Pot. (FAETP inf.)	kg DCB eq.	1.656E-04	1.652E-04	1.609E-04	1.649E-04	1.366E-04	1.449E-04	1.216E-04	1.361E-04	1.445E-04	1.211E-04	1.319E-04	1.402E-04	1.169E-04	1.361E-04	1.446E-04	1.211E-04
	IO2+ v2.1 - Aquatic ecotoxicity - Midpoint	kg TEG eq. to water	3.962E-01	3.942E-01	3.735E-01	3.933E-01	4.112E-01	4.743E-01	3.067E-01	4.092E-01	4.723E-01	3.047E-01	3.896E-01	4.516E-01	2.840E-01	4.094E-01	4.727E-01	3.049E-01
	ILCD Ecotoxicity freshwater midpoint (v1.09)	CTUe	3.394E-03	3.375E-03	3.222E-03	3.389E-03	3.139E-03	2.722E-03	2.587E-03	3.120E-03	2.703E-03	2.568E-03	2.967E-03	2.551E-03	2.415E-03	3.123E-03	2.710E-03	2.572E-03
	ReCiPe 2016 Midpoint (E) - Freshwater ecotoxicity	kg 1,4 DB eq.	6.575E-06	6.540E-06	6.355E-06	6.543E-06	4.847E-06	6.034E-06	4.888E-06	5.769E-06	5.998E-06	4.853E-06	5.584E-06	5.813E-06	4.668E-06	5.783E-06	6.017E-06	4.868E-06
	TRACI 2.1, Ecotoxicity	CTUe	1.952E-03	1.942E-03	1.885E-03	1.942E-03	1.459E-03	1.447E-03	1.470E-03	1.898E-03	1.437E-03	1.460E-03	1.816E-03	1.380E-03	1.383E-03	1.897E-03	1.442E-03	1.464E-03
Eutrophication Potential (EP)	CML2001 - Jan. 2016, Eutrophication Potential (EP)	kg PO4 eq.	3.812E-05	3.824E-05	3.640E-05	3.796E-05	3.391E-05	3.520E-05	3.007E-05	3.373E-05	3.503E-05	2.989E-05	3.189E-05	3.319E-05	2.806E-05	3.330E-05	3.460E-05	2.947E-05
	IO2+ v2.1 - Aquatic eutrophication - Midpoint	kg PO4 eq.	3.427E-07	3.350E-07	3.243E-07	3.402E-07	2.691E-07	4.867E-07	2.642E-07	2.613E-07	4.790E-07	2.564E-07	2.507E-07	4.684E-07	2.458E-07	2.675E-07	4.852E-07	2.626E-07
	ILCD Eutrophication freshwater midpoint (v1.09)	kg PO4 eq.	1.122E-07	1.102E-07	1.067E-07	1.116E-07	8.848E-08	1.601E-07	8.687E-08	8.595E-08	1.576E-07	8.434E-08	8.244E-08	1.541E-07	8.084E-08	8.796E-08	1.596E-07	8.635E-08
	ReCiPe 2016 Midpoint (H) - Freshwater Eutrophication	kg PO4 eq.	1.126E-07	1.101E-07	1.066E-07	1.118E-07	8.671E-08	1.600E-07	8.679E-08	8.585E-08	1.575E-07	8.426E-08	8.235E-08	1.540E-07	8.076E-08	8.787E-08	1.595E-07	8.628E-08
	TRACI 2.1, Eutrophication	kg N eq.	1.852E-05	1.848E-05	1.771E-05	1.845E-05	1.422E-05	1.911E-05	1.472E-05	1.740E-05	1.906E-05	1.468E-05	1.663E-05	1.830E-05	1.391E-05	1.722E-05	1.889E-05	1.450E-05
Human toxicity	CML2001 - Jan. 2016, Human Toxicity Potential (HTP inf.)	kg DCB eq.	4.752E-02	4.752E-02	4.746E-02	4.751E-02	3.386E-02	3.431E-02	3.334E-02	3.386E-02	3.430E-02	3.333E-02	3.380E-02	3.424E-02	3.328E-02	3.386E-02	3.430E-02	3.333E-02
	ILCD Human toxicity midpoint, cancer effects (v1.09)	CTUh	1.440E-10	1.430E-10	1.342E-10	1.424E-10	1.263E-10	1.250E-10	1.128E-10	1.253E-10	1.239E-10	1.118E-10	1.165E-10	1.152E-10	1.031E-10	1.253E-10	1.242E-10	1.119E-10
	ReCiPe 2016 Midpoint (E) - Human toxicity, cancer	kg 1,4-DB eq.	4.118E-04	4.026E-04	3.797E-04	4.067E-04	3.260E-04	3.952E-04	3.321E-04	3.704E-04	3.861E-04	3.229E-04	3.474E-04	3.631E-04	2.999E-04	3.764E-04	3.925E-04	3.289E-04
	TRACI 2.1, Human toxicity, cancer	CTUh	6.782E-11	6.764E-11	6.592E-11	6.753E-11	4.936E-11	5.456E-11	4.942E-11	5.252E-11	5.439E-11	4.925E-11	5.080E-11	5.267E-11	4.753E-11	5.251E-11	5.439E-11	4.924E-11
Photochem. Ozone Creation Potential (POCP)	CML2001 - Jan. 2016, Photochem. Ozone Creation Potential (POCP)	kg C2H4 eq. to air	2.494E-05	2.492E-05	2.704E-05	2.501E-05	1.478E-05	1.714E-05	1.416E-05	1.477E-05	1.712E-05	1.414E-05	1.688E-05	1.924E-05	1.626E-05	1.570E-05	1.806E-05	1.508E-05
	IO2+ v2.1 - Photochemical oxidation - Midpoint	kg C2H4 eq. to air	1.102E-05	1.101E-05	1.065E-05	1.097E-05	9.154E-06	9.430E-06	8.169E-06	9.140E-06	9.416E-06	8.155E-06	8.785E-06	9.061E-06	7.800E-06	9.106E-06	9.384E-06	8.120E-06
	ILCD - Photochemical ozone formation midpoint, human health (v1.09)	kg NMVOC eq.	3.129E-04	3.128E-04	3.005E-04	3.118E-04	2.279E-04	2.617E-04	2.358E-04	2.581E-04	2.616E-04	2.357E-04	2.459E-04	2.493E-04	2.234E-04	2.549E-04	2.583E-04	2.324E-04
	ReCiPe 2016 Midpoint (H) - Photochemical Ozone Formation, Ecosystems	kg NOx eq.	4.148E-02	4.140E-02	4.073E-02	4.136E-02	3.090E-02	6.428E-02	3.034E-02	4.859E-02	6.420E-02	3.026E-02	4.786E-02	6.353E-02	2.959E-02	4.853E-02	6.421E-02	3.026E-02
Ozone Layer Depletion Potential	CML2001 - Jan. 2016 Ozone Layer Depletion Potential (ODP, steady state)	kg R-11 eq.	3.739E-13	3.739E-13	3.730E-13	3.738E-13	3.662E-13	3.899E-13	3.660E-13	3.662E-13	3.898E-13	3.659E-13	3.654E-13	3.890E-13	3.651E-13	3.662E-13	3.898E-13	3.659E-13
	IO2+ v2.1 - Ozone layer depletion - Midpoint	kg R-11 eq.	3.739E-13	3.739E-13	3.730E-13	3.738E-13	3.662E-13	3.899E-13	3.660E-13	3.662E-13	3.898E-13	3.659E-13	3.654E-13	3.890E-13	3.651E-13	3.662E-13	3.898E-13	3.659E-13
	ILCD - Ozone depletion midpoint (v1.09)	kg CFC-11 eq.	3.744E-13	3.743E-13	3.735E-13	3.742E-13	3.666E-13	3.906E-13	3.663E-13	3.665E-13	3.905E-13	3.662E-13	3.657E-13	3.897E-13	3.654E-13	3.665E-13	3.905E-13	3.662E-13
	ReCiPe 2016 Midpoint (H) - Stratospheric Ozone Depletion	kg CFC-11 eq.	2.778E-08	2.780E-08	2.702E-08	2.772E-08	2.098E-08	3.510E-08	2.148E-08	3.257E-08	3.501E-08	2.139E-08	3.190E-08	3.434E-08	2.072E-08	3.240E-08	3.490E-08	2.127E-08
	TRACI 2.1, Ozone Depletion Air	kg CFC-11 eq.	3.959E-13	3.958E-13	3.950E-13	3.957E-13	3.879E-13	4.118E-13	3.879E-13	5.252E-11	4.117E-13	3.879E-13	3.873E-13	4.109E-13	3.870E-13	3.881E-13	4.117E-13	3.878E-13
Terrestrial Acidification	CML2001 - Jan. 2016, Acidification Potential (AP)	kg SO2 eq.	3.073E-04	3.066E-04	2.985E-04	3.073E-04	2.544E-04	2.575E-04	2.279E-04	2.537E-04	2.568E-04	2.272E-04	2.456E-04	2.487E-04	2.191E-04	2.520E-04	2.552E-04	2.255E-04
	IO2+ v2.1 - Terrestrial acidification/nutrification - Midpoint	kg SO2 eq. to air	1.609E-03	1.608E-03	1.526E-03	1.602E-03	1.377E-03	1.382E-03	1.242E-03	1.376E-03	1.381E-03	1.241E-03	1.294E-03	1.299E-03	1.159E-03	1.352E-03	1.357E-03	1.217E-03
	ILCD Acidification midpoint (v1.09)	Mole of H+ eq.	3.815E-04	3.813E-04	3.689E-04	3.804E-04	3.180E-04	3.168E-04	2.846E-04	3.178E-04	3.167E-04	2.844E-04	3.055E-04	3.044E-04	2.721E-04	3.145E-04	3.134E-04	2.811E-04
	ReCiPe 2016 Midpoint (E) - Terrestrial Acidification	kg SO2 eq.	2.367E-04	2.365E-04	2.301E-04	2.361E-04	1.707E-04	1.948E-04	1.749E-04	1.958E-04	1.947E-04	1.747E-04	1.893E-04	1.883E-04	1.683E-04	1.941E-04	1.931E-04	1.731E-04
	TRACI 2.1, Acidification	kg SO2 eq.	3.383E-04	3.381E-04	3.267E-04	3.373E-04	2.468E-04	2.857E-04	2.547E-04	2.830E-04	2.855E-04	2.545E-04	2.715E-04	2.741E-04	2.431E-04	2.798E-04	2.824E-04	2.514E-04
Terrestrial ecotoxicity	CML2001 - Jan. 2016, Terrestric Ecotoxicity Potential (TETP inf.)	kg DCB eq.	2.836E-04	2.835E-04	2.763E-04	2.826E-04	2.230E-04	2.172E-04	2.054E-04	2.229E-04	2.172E-04	2.054E-04	2.158E-04	2.100E-04	1.982E-04	2.224E-04	2.167E-04	2.048E-04
	ReCiPe 2016 Midpoint (H) - Terrestrial ecotoxicity	kg 1,4-DB eq.	4.793E-02	4.792E-02	4.747E-02	4.786E-02	3.401E-02	3.999E-02	3.402E-02	3.551E-02	3.997E-02	3.401E-02	3.506E-02	3.952E-02	3.356E-02	3.548E-02	3.994E-02	3.398E-02

4-0-0 (17)2F-0 2	5-0-0 (19)2S-0 2	6-0-0 (18)2B-0 2	1-1-1 (29)1F-1Cs 2	2-1-1 (31)1S-1Cs 2	3-1-1 (30)1B-1Cs 2	1-1-2 (32)1F-1S 2	2-1-2 (34)1S-1S 2	3-1-2 (33)1B-1S 2	4-1-0 (20)2F-0.5W 2.5	5-1-0 (22)2S-0.5W 2.5	6-1-0 (21)2B-0.5W 2.5	4-0-1 (23)2F-0.5Cs 2.5	5-0-1 (25)2S-0.5Cs 2.5	6-0-1 (24)2B-0.5Cs 2.5	4-0-2 (26)2F-0.5S 2.5	5-0-2 (28)2S-0.5S 2.5	6-0-2 (27)2B-0.5S 2.5	4-1-1 (35)2F-1Cs 3	5-1-1 (37)2S-1Cs 3	6-1-1 (36)2B-1Cs 3	4-1-2 (38)2F-1S 3
0.1337	0.1618	0.1342	0.1680	0.1732	0.1525	0.1694	0.1746	0.1539	0.1547	0.1616	0.1616	0.1529	0.1598	0.1322	0.1543	0.1612	0.1336	0.1527	0.1596	0.1321	0.1541
0.1331	0.1597	0.1340	0.1668	0.1716	0.1521	0.1681	0.1730	0.1535	0.1532	0.1597	0.1597	0.1513	0.1578	0.1320	0.1526	0.1591	0.1334	0.1512	0.1577	0.1317	0.1526
0.1333	0.1609	0.1339	0.1677	0.1725	0.1523	0.1691	0.1739	0.1537	0.1544	0.1608	0.1608	0.1525	0.1589	0.1319	0.1539	0.1603	0.1333	0.1524	0.1588	0.1319	0.1538
0.1321	0.1585	0.1327	0.1655	0.1704	0.1510	0.1669	0.1717	0.1523	0.1520	0.1585	0.1585	0.1501	0.1566	0.1307	0.1515	0.1579	0.1321	0.1501	0.1565	0.1307	0.1514
0.1333	0.1608	0.1338	0.1677	0.1725	0.1523	0.1691	0.1739	0.1537	0.1544	0.1607	0.1607	0.1524	0.1588	0.1319	0.1538	0.1602	0.1333	0.1524	0.1587	0.1318	0.1538
1.058E-04	1.381E-04	1.069E-04	1.314E-04	1.398E-04	1.164E-04	1.357E-04	1.441E-04	1.207E-04	1.264E-04	1.377E-04	1.377E-04	1.222E-04	1.334E-04	1.022E-04	1.264E-04	1.378E-04	1.065E-04	1.217E-04	1.330E-04	1.018E-04	1.260E-04
2.736E-01	5.004E-01	2.769E-01	3.866E-01	4.496E-01	2.820E-01	4.074E-01	4.707E-01	3.029E-01	4.142E-01	4.984E-01	4.984E-01	3.936E-01	4.778E-01	2.542E-01	4.144E-01	4.990E-01	2.751E-01	3.916E-01	4.758E-01	2.522E-01	4.124E-01
2.269E-03	2.500E-03	2.318E-03	2.948E-03	2.532E-03	2.396E-03	3.105E-03	2.691E-03	2.553E-03	3.035E-03	2.481E-03	2.481E-03	2.882E-03	2.329E-03	2.146E-03	3.038E-03	2.490E-03	2.303E-03	2.863E-03	2.310E-03	2.127E-03	3.020E-03
4.271E-06	5.855E-06	4.326E-06	5.548E-06	5.778E-06	4.633E-06	5.748E-06	5.981E-06	4.832E-06	5.511E-06	5.820E-06	5.820E-06	5.326E-06	5.633E-06	4.106E-06	5.526E-06	5.841E-06	4.305E-06	5.291E-06	5.600E-06	4.070E-06	5.491E-06
1.294E-03	1.280E-03	1.310E-03	1.806E-03	1.350E-03	1.373E-03	1.887E-03	1.432E-03	1.454E-03	1.877E-03	1.270E-03	1.270E-03	1.800E-03	1.198E-03	1.223E-03	1.880E-03	1.275E-03	1.303E-03	1.790E-03	1.183E-03	1.213E-03	1.870E-03
2.669E-05	3.413E-05	2.729E-05	3.171E-05	3.301E-05	2.788E-05	3.313E-05	3.443E-05	2.929E-05	3.222E-05	3.396E-05	3.396E-05	3.039E-05	3.212E-05	2.528E-05	3.180E-05	3.353E-05	2.669E-05	3.021E-05	3.194E-05	2.510E-05	3.162E-05
2.377E-07	5.348E-07	2.380E-07	2.429E-07	4.606E-07	2.381E-07	2.597E-07	4.775E-07	2.549E-07	2.368E-07	5.270E-07	5.270E-07	2.261E-07	5.164E-07	2.196E-07	2.429E-07	5.332E-07	2.364E-07	2.184E-07	5.086E-07	2.119E-07	2.352E-07
7.814E-08	1.760E-07	7.825E-08	7.991E-08	1.516E-07	7.831E-08	8.543E-08	1.571E-07	8.382E-08	7.786E-08	1.734E-07	1.734E-07	7.436E-08	1.699E-07	7.221E-08	7.987E-08	1.755E-07	7.773E-08	7.183E-08	1.674E-07	6.968E-08	7.735E-08
7.807E-08	1.758E-07	7.818E-08	7.982E-08	1.514E-07	7.823E-08	8.534E-08	1.570E-07	8.375E-08	7.777E-08	1.733E-07	1.733E-07	7.427E-08	1.698E-07	7.215E-08	7.978E-08	1.753E-07	7.766E-08	7.174E-08	1.672E-07	6.962E-08	7.725E-08
1.324E-05	1.930E-05	1.345E-05	1.659E-05	1.826E-05	1.387E-05	1.718E-05	1.885E-05	1.446E-05	1.703E-05	1.926E-05	1.926E-05	1.627E-05	1.850E-05	1.265E-05	1.686E-05	1.909E-05	1.324E-05	1.623E-05	1.846E-05	1.261E-05	1.682E-05
2.860E-02	2.990E-02	2.861E-02	3.380E-02	3.424E-02	3.327E-02	3.385E-02	3.430E-02	3.333E-02	2.931E-02	2.990E-02	2.990E-02	2.925E-02	2.984E-02	2.855E-02	2.931E-02	2.990E-02	2.860E-02	2.925E-02	2.984E-02	2.854E-02	2.930E-02
9.920E-11	1.187E-10	1.025E-10	1.155E-10	1.142E-10	1.021E-10	1.243E-10	1.232E-10	1.109E-10	1.193E-10	1.177E-10	1.177E-10	1.106E-10	1.090E-10	9.269E-11	1.194E-10	1.181E-10	1.015E-10	1.096E-10	1.079E-10	9.167E-11	1.184E-10
2.974E-04	3.900E-04	3.055E-04	3.383E-04	3.539E-04	2.907E-04	3.672E-04	3.833E-04	3.197E-04	3.597E-04	3.809E-04	3.809E-04	3.367E-04	3.579E-04	2.733E-04	3.657E-04	3.877E-04	3.023E-04	3.275E-04	3.487E-04	2.641E-04	3.565E-04
4.321E-11	5.015E-11	4.329E-11	5.063E-11	5.250E-11	4.736E-11	5.234E-11	5.422E-11	4.907E-11	4.748E-11	4.998E-11	4.998E-11	4.576E-11	4.826E-11	4.139E-11	4.747E-11	4.998E-11	4.311E-11	4.559E-11	4.809E-11	4.122E-11	4.730E-11
1.153E-05	1.454E-05	1.057E-05	1.686E-05	1.922E-05	1.624E-05	1.568E-05	1.804E-05	1.506E-05	1.138E-05	1.452E-05	1.452E-05	1.350E-05	1.664E-05	1.267E-05	1.231E-05	1.546E-05	1.098E-05	1.348E-05	1.662E-05	1.265E-05	1.230E-05
7.159E-06	8.902E-06	7.218E-06	8.771E-06	9.047E-06	7.786E-06	9.092E-06	9.370E-06	8.106E-06	8.518E-06	8.888E-06	8.888E-06	8.163E-06	8.533E-06	6.849E-06	8.484E-06	8.858E-06	7.170E-06	8.149E-06	8.519E-06	6.835E-06	8.470E-06
2.068E-04	2.447E-04	2.102E-04	2.457E-04	2.492E-04	2.233E-04	2.547E-04	2.582E-04	2.323E-04	2.399E-04	2.446E-04	2.446E-04	2.277E-04	2.323E-04	1.977E-04	2.366E-04	2.413E-04	2.067E-04	2.275E-04	2.321E-04	1.976E-04	2.365E-04
2.659E-02	7.189E-02	2.663E-02	4.777E-02	6.345E-02	2.951E-02	4.845E-02	6.413E-02	3.018E-02	5.090E-02	7.180E-02	7.180E-02	5.023E-02	7.114E-02	2.588E-02	5.091E-02	7.181E-02	2.655E-02	5.015E-02	7.105E-02	2.579E-02	5.082E-02
3.633E-13	3.952E-13	3.633E-13	3.653E-13	3.889E-13	3.650E-13	3.661E-13	3.897E-13	3.658E-13	3.636E-13	3.951E-13	3.951E-13	3.628E-13	3.943E-13	3.625E-13	3.636E-13	3.951E-13	3.633E-13	3.627E-13	3.942E-13	3.624E-13	3.635E-13
3.633E-13	3.952E-13	3.633E-13	3.653E-13	3.889E-13	3.650E-13	3.661E-13	3.897E-13	3.658E-13	3.636E-13	3.951E-13	3.633E-13	3.628E-13	3.943E-13	3.625E-13	3.636E-13	3.951E-13	3.633E-13	3.627E-13	3.942E-13	3.624E-13	3.635E-13
3.636E-13	3.960E-13	3.636E-13	3.656E-13	3.896E-13	3.653E-13	3.664E-13	3.904E-13	3.661E-13	3.639E-13	3.959E-13	3.959E-13	3.631E-13	3.951E-13	3.627E-13	3.639E-13	3.959E-13	3.635E-13	3.630E-13	3.950E-13	3.627E-13	3.638E-13
1.912E-08	3.751E-08	1.934E-08	3.180E-08	3.425E-08	2.062E-08	3.236E-08	3.481E-08	2.118E-08	3.416E-08	3.742E-08	3.742E-08	3.349E-08	3.675E-08	1.858E-08	3.405E-08	3.730E-08	1.914E-08	3.340E-08	3.665E-08	1.849E-08	3.395E-08
3.853E-13	4.171E-13	3.853E-13	3.872E-13	4.109E-13	3.870E-13	3.880E-13	4.117E-13	3.878E-13	3.856E-13	4.171E-13	4.171E-13	3.847E-13	4.162E-13	3.844E-13	3.855E-13	4.170E-13	3.852E-13	3.847E-13	4.162E-13	3.843E-13	3.855E-13
1.989E-04	2.410E-04	2.014E-04	2.449E-04	2.480E-04	2.183E-04	2.513E-04	2.544E-04	2.248E-04	2.361E-04	2.403E-04	2.403E-04	2.280E-04	2.321E-04	1.926E-04	2.344E-04	2.386E-04	1.990E-04	2.273E-04	2.314E-04	1.919E-04	2.337E-04
1.094E-03	1.306E-03	1.120E-03	1.293E-03	1.298E-03	1.158E-03	1.351E-03	1.356E-03	1.216E-03	1.299E-03	1.305E-03	1.305E-03	1.217E-03	1.223E-03	1.037E-03	1.275E-03	1.281E-03	1.095E-03	1.216E-03	1.222E-03	1.036E-03	1.274E-03
2.460E-04	2.954E-04	2.523E-04	3.053E-04	3.042E-04	2.719E-04	3.143E-04	3.132E-04	2.809E-04	2.967E-04	2.952E-04	2.952E-04	2.843E-04	2.829E-04	2.398E-04	2.933E-04	2.919E-04	2.488E-04	2.841E-04	2.827E-04	2.396E-04	2.931E-04
1.524E-04	1.809E-04	1.543E-04	1.892E-04	1.882E-04	1.682E-04	1.940E-04	1.930E-04	1.730E-04	1.822E-04	1.808E-04	1.808E-04	1.765E-04	1.736E-04	1.477E-04	1.805E-04	1.792E-04	1.525E-04	1.756E-04	1.742E-04	1.476E-04	1.804E-04
2.235E-04	2.682E-04	2.268E-04	2.714E-04	2.739E-04	2.429E-04	2.797E-04	2.822E-04	2.512E-04	2.646E-04	2.680E-04	2.680E-04	2.532E-04	2.566E-04	2.152E-04	2.615E-04	2.649E-04	2.235E-04	2.530E-04	2.564E-04	2.150E-04	2.613E-04
1.793E-04	1.951E-04	1.794E-04	2.157E-04	2.099E-04	1.981E-04	2.223E-04	2.166E-04	2.048E-04	2.028E-04	1.951E-04	1.951E-04	1.956E-04	1.879E-04	1.721E-04	2.022E-04	1.946E-04	1.788E-04	1.955E-04	1.878E-04	1.721E-04	2.022E-04
2.937E-02	3.734E-02	2.939E-02	3.505E-02	3.951E-02	3.354E-02	3.547E-02	3.993E-02	3.396E-02	3.138E-02	3.733E-02	3.733E-02	3.092E-02	3.687E-02	2.892E-02	3.135E-02	3.730E-02	2.934E-02	3.091E-02	3.686E-02	2.890E-02	3.133E-02

The model also provides a number of alternatives to achieve credit points for the major materials, which are regulated in Green Star Credits 13 – "Indoor Pollutants", 19 – "Life-cycle Impacts", and 20 – "Responsible Building Material" regarding the analysis for life-cycle greenhouse gas emissions of timber, steel, and concrete. Credits 13 and 20, related to timber product assessment, help the project achieve 2 available points when the project employs more than 95% of reused wood products (Green Building Council of Australia, 2015a). Up to 2 points can be obtained for the construction by satisfying the section in Credit 19 – Life-cycle impacts – Steel, which illustrates a reduction in the quantity of reinforcement bars in comparison with the standard practice. These credits are awarded to the project when the designing team, in their quantity breakdown report, demonstrates the deduction of reinforcement quantity compared to the standard level.

Three points can be obtained when the engineering team uses methods of Credit 19 to reduce the environmental impact of concrete by replacing Portland cement with supplementary cementitious materials, reclaiming 50% the amount of water, and using alternative coarse or fine aggregates. The remaining six points available mentioned in the model are granted to the project when the project's life-cycle impact is assessed with six major impact factors, namely, stratospheric ozone depletion potential, acidification potential of land and water, eutrophication potential, tropospheric ozone formation potential, photochemical ozone creation potential (POCP ethylene equivalents), and mineral and fossil fuel depletion (abiotic depletion), and five additional impact categories: human toxicity, land use, resource depletion – water, ionizing radiation, and particulate matter.

Apart from the significant reduction of the building's negative impact to the environment by using the suggested alternatives described in the chapter, the maximum points that the model can help the project achieve is 34 points to add to the Green Star certificate.

5 Life-cycle greenhouse gas emissions

Case studies and validation

5.1 Introduction

This chapter focuses on the life-cycle greenhouse gas emissions calculations for the case studies that are Green Star-certified. The chapter is divided into two main parts. The first part describes the estimation scope, functional unit, and system boundary of the case studies to be assessed by the life-cycle greenhouse gas emissions assessment model. Different validation methods are discussed and chosen to confirm that the analytical model deployed is sustainable and suitable for the model's intended purpose. This chapter comprehensively discusses the validation case studies as well as validating methods.

In the second part, the environmental impacts and the credit criteria in terms of life-cycle greenhouse gas emissions are discussed. As mentioned in the previous chapters, materials used for building structures play a significant role in life-cycle greenhouse gas emissions management. It is necessary to compare case studies with studies on reference buildings, as well as to apply materials options that match the National Construction Code of Australia (NCC) requirements, to validate the model's results. The certified Green Star projects are selected to consider and compare the credits achieved related to the sources of greenhouse gas emissions with the credits selected by the life-cycle greenhouse gas emissions model. The complexity of this studied model lies in the ability to conduct analysis of a part of building envelopes or materials used for construction projects. With the flexibility to choose the parameters in the model, the study will open feasible directions that can adapt to other green building standards in the world.

5.2 Case studies demonstrating greenhouse gas emission reduction strategies

The purpose of the studied model is to demonstrate how life-cycle greenhouse gas emissions assessment can be applied when deploying major materials such as timber, concrete, and steel and comparing the building's fabric elements between reference buildings and the case studies. The assessment also highlights the guidelines to achieve Green Star credit points related to the reduction of greenhouse gas emissions.

The model used to assess life-cycle greenhouse gas emissions contains a series of estimation sequences. The accurateness of the alternatives depends on the reference building attributes, the building design, and the accuracy of the work breakdown estimation. Analyzing the life-cycle greenhouse gas emissions by the unit analysis approach helps designers by providing them with a flexible tool to estimate the environmental impact for the whole building after they estimate the project's breakdown quantities.

5.2.1 *Scope of validation*

The amount of life-cycle greenhouse gas emissions produced from commercial office building's alternatives is calculated to determine the lowest level of greenhouse gas emissions and associated environmental impacts of the buildings' materials and fabrics.

The selection of impact categories of the book complies with the guidance in the credit: life-cycle impact in the Green Star – Design & As Built rating tool, including global warming potential, ozone depletion potential, photochemical ozone creation potential, eutrophication potential, aquatic ecotoxicity, terrestrial ecotoxicity, acidification potential, and human toxicity potential. Life-cycle methodologies to use in the model are ReCiPe 2016, ILCD, TRACI 2.1, and I02+ v2.1, which can be analyzed in GaBi.

5.2.2 *Functional unit*

The case studies used for the model validation are Green Star-certified projects. The studies used the Building Code of Australia deemed-to-satisfy approach to identify the building's envelopes. The three studied buildings are 5-star rated commercial buildings. One is located in Canberra, ACT, with 1,200 m^2, and the others are located in South Melbourne and Perth. These buildings have 14,230 m^2, 40,020 m^2, and 45,000 m^2 of gross floor area, respectively.

The first case study to use for model validation is located in Canberra, ACT. This case study is a three-storey commercial building with a lettable area of 14,230 m^2. Several initiatives are implemented to achieve the Green Star point rating, and these initiatives satisfy major key criteria of the rating system. Green Star building performance in the material improvement of the building satisfied 80% of the requirements in the material criteria of the rating tool. The project was carried out in compliance with the New South Wales Environment Management System Guidelines 1998 and the best practice Australian Government Guidelines.

This building uses a building management system to monitor and control energy consumption within the building. The system helps analyze the energy use and provide an energy-saving strategy for the future. The concrete structure in this building replaces over 40% of cement by a silica fume product and increases the strength of the concrete in the range of 25–100 MPa as well. Over 80% of the construction waste in this project is collected, recycled, and reused.

The second case study is located in Carlton, Victoria. The building was designed to be a model of top-quality environmental performance that established the Green Building Partnership and Australian Conservation Foundation's commitment to environmentally sustainable development. The building reduces approximately 70% energy use in comparison with the average office building of the same size in the Melbourne area.

Natural ventilation and lighting improve energy performance to one-third of the quantity used in similarly sized buildings. Solar panels are used to supplement the main energy supply. Green power is sourced as the main energy supply. A three-step water recycling system supplies potable and non-potable water to tenants. This building uses an average of 60% recycled crushed concrete. The supplementary cementitious material used in the concrete in this building was fly ash.

The third case study building is located in Perth, Western Australia. This is a complex with gross office floor area of 45,000 m^2. This building was designed with the intention to harmonize sustainability principles into the process of the building. With various initiatives, the building targeted toward a zero-emission building by using rammed earth and recycled materials for thermal mass walls, deploying a solar energy collection and thermal management system, and using recycled materials in the construction phase to reduce landfill waste. The wall structure of this building is 220 cm thickness wall to balance inside versus outdoor temperature. The building management system controls the temperature within the office space.

This project performed many initiatives to use recycled materials when possible or to reduce construction waste to landfill as much as possible. The thermal mass walls are built with a mixture of 10% cement and stabilized building mortar, which are made of recycled brick and rammed recycled earth. The building walls are used with the compressive strength range of 25–100 MPa of concrete and used 40% of fly ash to replace cement in concrete. Slab concrete is made of fly ash, demolished concrete, and glass waste. This reduced the amount of cement and aggregates used as well as the waste materials for the landfill. The industrial products are used from local sources to reduce the energy during the construction phase.

Many advantage features are applied to the projects to obtain Green Star ratings. The global warming potential impact was estimated for these buildings to be 442 kg CO2-eq/m^2, 420 kg CO2-eq/m^2, and 417 kg CO2-eq/m^2 for Case 1, 2, and 3, respectively.

Table 5.1 Information about case studies

Case study	*Case 1*	*Case 2*	*Case 3*
Location	Canberra, ACT	South Melbourne, Carlton, Victoria	Perth, Western Australia
Gross floor area, m^2	14,230	40,020	45,000
Green Star rating	5-star	5-star	5-star
GWP per square area (kg CO2-eq/m^2)	290	380	440

Building envelope characteristics of the case studies and reference building are introduced in Table 5.2. These features offer all primary key characteristics to satisfy the credit criteria. The main materials used for envelopes of the buildings are listed in the table. The total environmental impact produced from production, construction, maintenance, and demolition phases is determined by eight impact factors mentioned in Table 3.8 across the project's predicted life-cycle of 100 years (Green Building Council of Australia, 2015a). The envelope system is described in the Australian Building Codes Board (2016) by six elements usually denoting outdoor air film, water-proof membrane, concrete slab, air space, plaster-board gypsum, and indoor air film.

Table 5.3 shows the reference building structure of a metal roof, masonry wall, and concrete floor. The wall structure of this building is 110 mm clay brickwork masonry. In this building, concrete slabs with 150 mm thickness are used to construct floors. The roof system of the case study 1 has the structure comprising a steel roof frame, metal roof cladding, double-sided antiglare foil, and 10 mm plasterboard. The wall structure of this building uses a hollow concrete block as the primary material with 10 mm plasterboard and wall batts R2.7 (90 mm). Same with the reference building, the structure of the floor is concrete slab structure with 150 mm thickness and a fibrous insulation board R3.0. Table 5.2 illustrates the detailed structure, with the metal roof made up of a metal sheet with 0° –5° angle pitch over a steel frame and plasterboard gypsum.

The roof structure of both case studies 2 and 3 are built with a flat concrete roof structure, plasterboard ceiling, blanket thermal resistance, and foil-faced Rm applied to the ceiling. Total R-value of these structures is measured based on all elements of the structure including main building materials and outdoor and indoor air films. The wall structure in case studies 2 and 3 is the structure of masonry brick wall and 10 mm plasterboard. The floors use a concrete slab of 150 mm and a double-sided antiglare board. R-value data of elements for the structures collected from the model are summarized in Table 5.4.

The next part of the model is to validate how a commercial building can achieve Green Star points in Credit 14 and 19, which are related to timber, steel, and concrete use in the buildings. Moreover, the next part describes how life-cycle greenhouse gas emissions can be assessed when choosing original or supplemental cementitious material concrete types and aggregate types following the credit's requirements.

The assessment also highlights the environmental impact of alternative aggregates or water content in 1 m^3 of concrete. The content of cement and aggregate replacement is complied with Green Star requirements to achieve the credit points related to life-cycle impact of concrete use in commercial buildings. To assess the credit points achieved related to materials use in the buildings, the research collects the quantity breakdown data of these case studies (Table 5.5 and Table 5.6).

The material for the case study 1 structure is made of concrete with 60% cement and 40% silica fume replacement. With the same ratio of cement reduction, the other two case studies use fly ash as supplementary cementitious materials. While

Table 5.2 Detailed description of building elements
(Australian Building Codes Board, 2016)

Structure	*Structure description*	*Structure*	*Structure description*
Metal roof	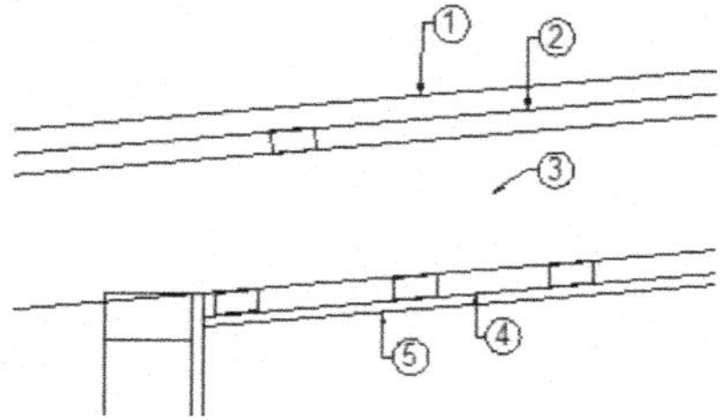	Masonry wall	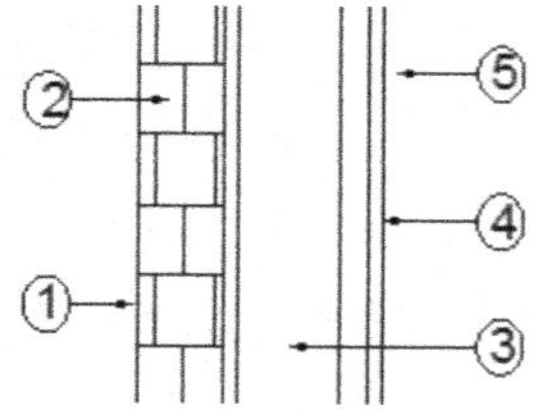
Concrete roof	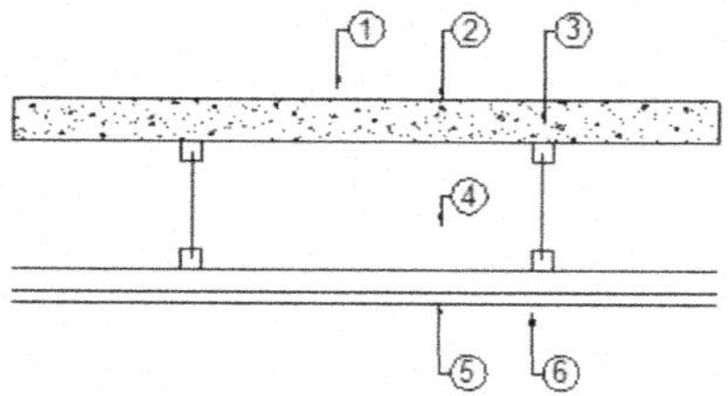	Concrete wall	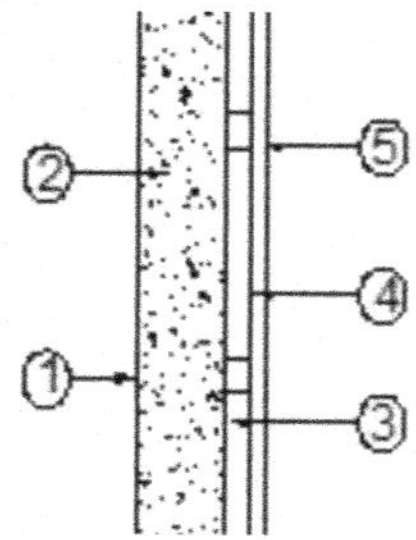
Concrete slab	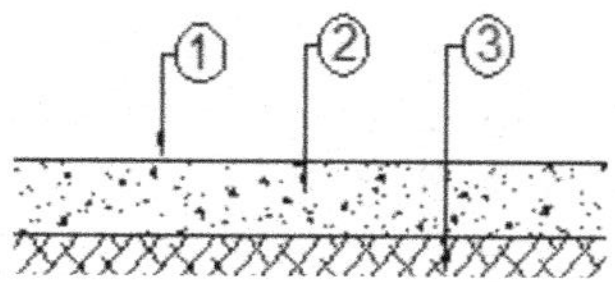		

Table 5.3 Characteristics of building elements used in the case studies versus reference building elements

Element	*Reference building*	*Case 1*	*Case 2*	*Case 3*
Roof	Metal roof	Steel roof frame, 10 mm plasterboard, membrane roof systems with double-sided antiglare foil	Flat concrete slab 150 mm, 10 mm plasterboard	Flat concrete slab 150 mm, plasterboard, foil-faced R3.3 blanket, 10 mm plasterboard
Walls	Masonry wall 110 mm	Concrete block with aluminum cladding panel, 10 mm plasterboard, wall batts R2.7 (90 mm)	Masonry brick wall 110 mm, 10 mm plasterboard, stud wall batts R1.0 (50 mm)	Masonry brick wall 2 × 110 mm, 10 mm plasterboard, stud wall batts R2.7 (90 mm)
Floor	Concrete slab	Concrete slab, 150 mm thickness, with fibrous insulation board R3.0	Concrete slab, 150 mm thickness, with double-sided antiglare EPS board R1.5	Concrete slab, 150 mm thickness, with PIR rigid board system R5.5 (120 mm)

crushed slag is replaced with coarse aggregate in concrete in case study 1, manufactured sand is used for case studies 2 and 3. The water for all concrete mixing used in these case studies is 50% reclaimed water.

The quantity data of timber and steel for the three case studies are shown in Table 5.6. Life-cycle greenhouse gas emissions of these materials are assessed by the model in Section 5.3 of this chapter.

5.2.3 System boundaries

All data of life-cycle greenhouse gas emissions linked with the project including raw material production, manufacture, construction, operation, demolition, disposal, and recycling of materials are determined. The elements of the building in the model assessment included roof, floors, external walls, and finishes.

For the credits associated with a material's life-cycle assessment, the study investigates the life-cycle of concrete, which is typically used for building in Australia with strength in the strength range of 20–100 MPa. Supplementary cementitious materials used in the model are fly ash, silica fume, and blast furnace slag with the cement content replacement in the ratio of 30–40%. Coarse and fine-aggregates also are considered for use as alternative aggregates with the ratio of 40% of crushed slag aggregate or 25% of manufactured sand. In addition, the model demonstrates the impact of the reduction of water to the amount of greenhouse gas emissions as well as Green Star point contribution to the project.

Table 5.4 Information of case studies and reference building envelope's R-value

Project	*Roof*		*Wall*		*Floor*	
Reference building	Metal roof cladding	0.00	Outdoor air film	0.04	Interior air film	0.16
	Outdoor air film	0.04	110 mm brickwork	0.18	Concrete slab (150 mm)	0.1
	Indoor air film	0.11	Indoor air film	0.61	Ground thermal resistance	0.58
Case study 1	Outdoor air film	0.04	Outdoor air film	0.04	Interior air film	0.16
	Metal roof cladding	0.00	Hollow concrete block	0.20	Concrete slab (150 mm)	0.1
	Double-sided antiglare foil	1.01	Air gap 20 mm	2.1	Subfloor air film	0.16
	10 mm plasterboard	0.06	10 mm plasterboard	0.06	Fibrous insulation board R3.0	4.1
	100 mm spacer and air gap	0.43	Wall batts R2.7 (90 mm)	3.8	Ground thermal resistance	0.58
	Indoor air film	0.11	Indoor air film	0.12		
Case study 2	Outdoor air film	0.04	Outdoor air film	0.04	Interior air film	0.16
	Foil-faced R1.3 blanket	1.37	110 mm brickwork	0.18	Concrete slab (150 mm)	0.1
	Concrete slab (150 mm)	0.1	Air gap 20 mm	2.1	25 mm air gap	1.8
	Non-ventilated air gap	2.23	10 mm plasterboard	0.06	Double-sided antiglare EPS board R1.5	4.4
	Ceiling insulation R1.2 (50 mm)	1.2	Indoor air film	0.12		
	10 mm plasterboard	0.06	Stud wall batts R1.5 (75 mm)	2.5	Ground thermal resistance	0.58
	25 mm spacer and air gap	0.43				
	Indoor air film	0.11				
Case study 3	Outdoor air film	0.04	Outdoor air film	0.04	Interior air film	0.16
	Blanket thermal resistance, foil-faced R3.3	3.48	2 × 110 mm brickwork	0.36	Concrete slab (150 mm)	0.1
			Air gap 20 mm	2.1	Subfloor air film	0.16
	Non-ventilated air gap	2.23	10 mm plasterboard	0.06	PIR board	7.1
	Concrete slab (150 mm)	0.1	Indoor air film	0.12	Ground thermal resistance	0.58
	10 mm plasterboard	0.06	Stud wall batts R2.7 (90 mm)	3.8		
	Air gap 20 mm	0.43				
	Indoor air film	0.11				

Table 5.5 Concrete quantity used in the case studies

Case study	*Case 1*	*Case 2*	*Case 3*
20 MPa (kg)	136.65	384.25	576.1
25 MPa (kg)	205	576.5	864.3
32 MPa (kg)	307.5	864.75	1,296.5
40 MPa (kg)	375.75	1,056.75	1,584.3
50 MPa (kg)	854	2,401.75	3,600.8
65 MPa (kg)	956.5	2,690	4,032.95
80 MPa (kg)	273.25	768.5	1,152.15
100 MPa (kg)	307.5	864.75	1,296.5
SCM	40% Silica fume	40% Fly ash	40% Fly ash
50% water reclaiming	☒	☒	☒
Alternative fine aggregate	Crushed slag aggregate	Manufactured sand	Manufactured sand

Table 5.6 Timber and steel quantity used in the case studies

Case study	*Case 1*	*Case 2*	*Case 3*
Timber	503.99	1,417.41	2,125.05
Steel quantity	706.15	1,985.95	2,977.44

5.3 Discussion on life-cycle greenhouse gas emissions assessment and credit satisfaction

Getting points in the pathways of the credits related to greenhouse gas emissions reduction helps the buildings reach the required green points level in Green Star. The comparison of environmental impact results in the case study assessment is implemented with CML 2001, ReCiPe, ILCD, TRACI, and Impact +02 methodologies, which are the methods used to estimate life-cycle impact per square meter of the buildings.

The credits for evaluation of environmental impact were achieved by all the case study buildings and lead to the results similar to the simulation. The environmental impacts used to estimate for these case studies are focused on global warming potential, ozone layer depletion potential, photochemical ozone creation potential, eutrophication potential, aquatic ecotoxicity, terrestrial ecotoxicity, acidification potential, human toxicity potential. The results of global warming potential impact show that the environmental impact increased along with the change in energy consumption when the building envelope improved in relation to the materials in the reference building (Table 5.7).

The model provides the ability to compare the environmental impact of the case studies and the reference building. Global warming potential results are fairly similar among the studied methodologies (Figure 5.1). The reference building and case study 1 results show that buildings with metal structures consume

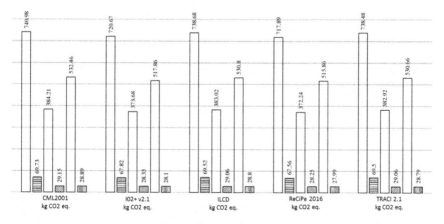

Table 5.7 Location-wise estimation of annual energy consumption for case studies and reference building (MJ/m²)

Location	*Reference*	*Case study*
Canberra-Queanbeyan	3,020.552	276.229
South Melbourne, Carlton, Victoria	1,566.218	118.8432
Perth	2,170.527	117.7657

more energy to keep the comfortable indoor temperature than concrete and masonry structures. Moreover, the thermal insulation materials play a crucial role in the building performance to reduce the environmental impact, as illustrated in Figure 5.1 (Papadopoulos, 2005). All buildings in the case studies that applied the insulating materials have significantly reduced the impact of all considered environmental aspects.

As seen in Figures 5.2 and 5.3, the ecotoxicity and human-toxicological impacts are calculated for potential leaks into groundwater by the different methods, leading to the diverse impact results. The resulting aquatic ecotoxicity (triethylene glycol equivalents – TEG eq) using the I02+ method for the three consecutive case studies are 397.16, 166.04, and 164.54 TEG-eq (Vandecasteele et al., 2015). Although both ILCD and TRACI 2.1 methods use comparative toxic unit – CTUe for aquatic ecotoxicity and comparative toxic unit for human – CTUh for human-toxicological impact, there is noticeable dissimilarity between the effect results (Vahidi & Zhao, 2017).

Electricity consumption plays a crucial role in all three case studies in terms of the human-toxicological impact. The substance results from these structures provide evidence that energy consumption needs be reduced to mitigate the human

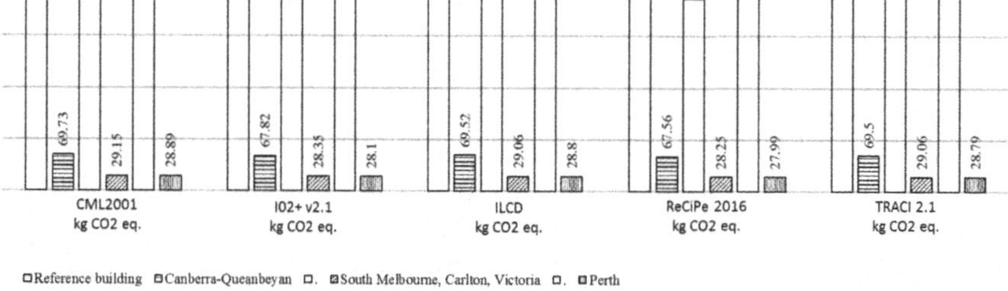

Figure 5.1 Global warming potential impact

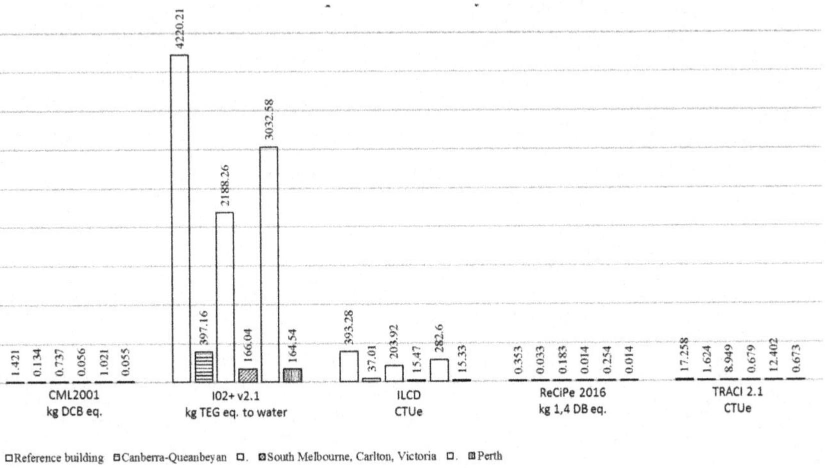

Figure 5.2 Aquatic ecotoxicity impact

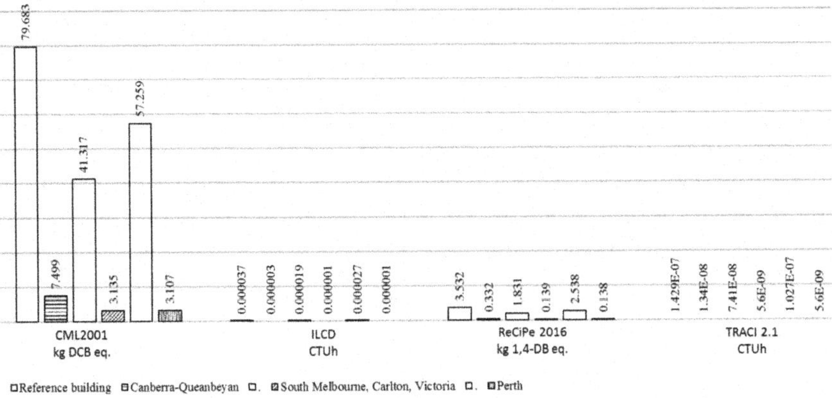

Figure 5.3 Human-toxicological impact

health risk. While ReCiPe 2016 deploys an average international factor based on specified location factors, the impact on human health is interpreted by the ILCD method by the LOTOS-EUROS model and the I02+ method by Eco-Indicator 99. By deploying a concrete roof and floor structure, the 110 mm masonry wall in case study 2, and the 220 mm masonry wall in the case study 3, along with effective thermal insulating layers, the environmental impacts are significantly reduced in these case studies.

The eutrophication potential results followed by their respective uncertainties obtained by CML 2001, I02+, ILCD, and ReCiPe methods are analyzed in terms of kg PO4 eq and the TRACI 2.1 method in terms of kg N-eq. The values

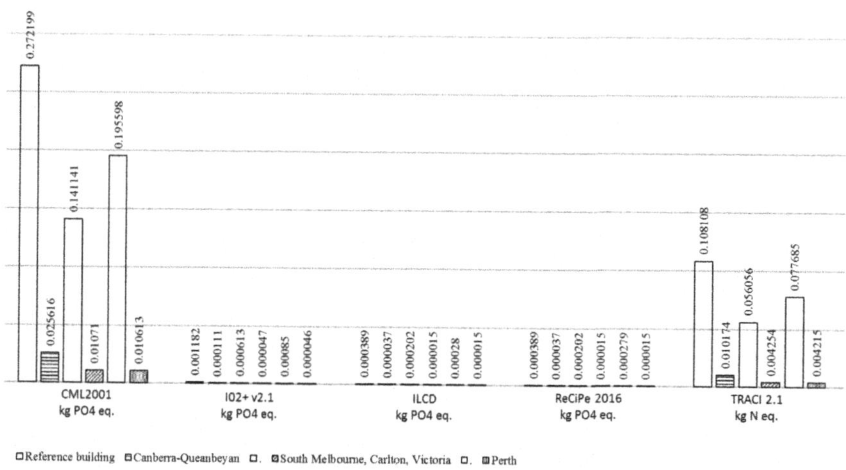

Figure 5.4 Eutrophication potential impact

obtained by I02+, ILCD, and ReCiPe are not significant, in comparison to the ones obtained by the other two methods (Figure 5.4) (Heijungs et al., 1992; Zelm, 2009; Bare, 2012). Figure 5.5 shows the results of photochemical ozone creation potential, by the CML 2001, I02+, ILCD, and ReCiPe methodologies, differing due to different factors employed by each method.

With regard to the impact on ozone layer depletion potential, the impact of brick is significant as well. Breakdown of the analysis of the structure using brick masonry shows that the impact on ozone layer depletion potential of this structure is higher than for the other envelopes. Choosing a method to evaluate the impact also plays a significant role in the entire assessment process. The primary substance that contributes to the ozone depletion potential impact is different

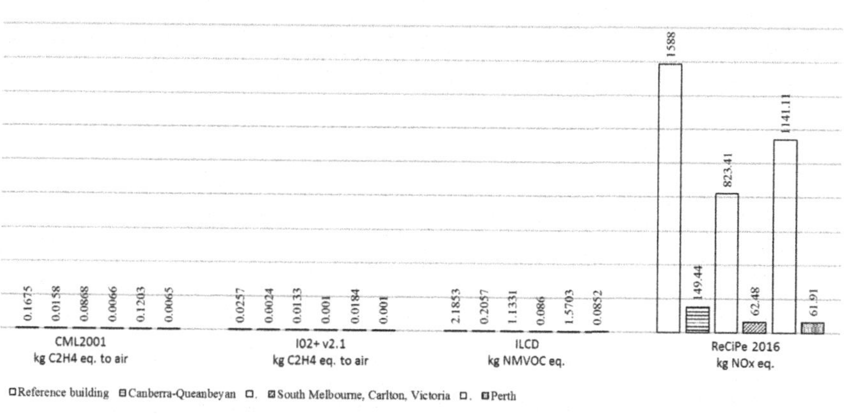

Figure 5.5 Photochemical ozone creation potential impacts

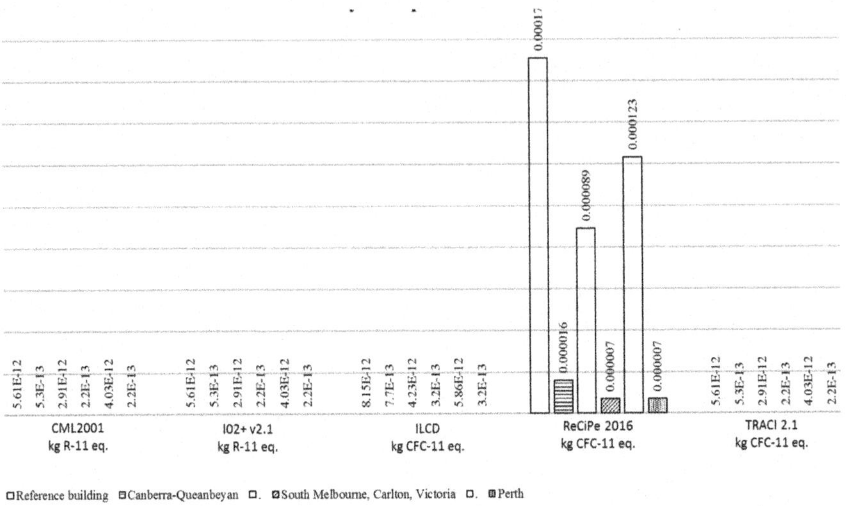

Figure 5.6 Ozone layer depletion impact

depending on applied assessment methods. When using the ReCiPe method, nitrous oxide (N2O) is the primary substance contributing to this category, while carbon tetrachloride (CCl4) scores higher impact than the other substances when applying the other three methods: ILCD, IMPACT 02, and TRACI.

In the analysis of the acidification potential impact factor, the patterns of the model's results are identical. The weight of acid emissions is expressed with the reference unit of kg SO2-equivalence, which is evaluated by the ReCiPe, TRACI, and I0+2 methodologies, and the reference unit of a mole of H+ equivalent by the ILCD method. In ReCiPe and I0+2 impact analysis, the brick structure also contributes

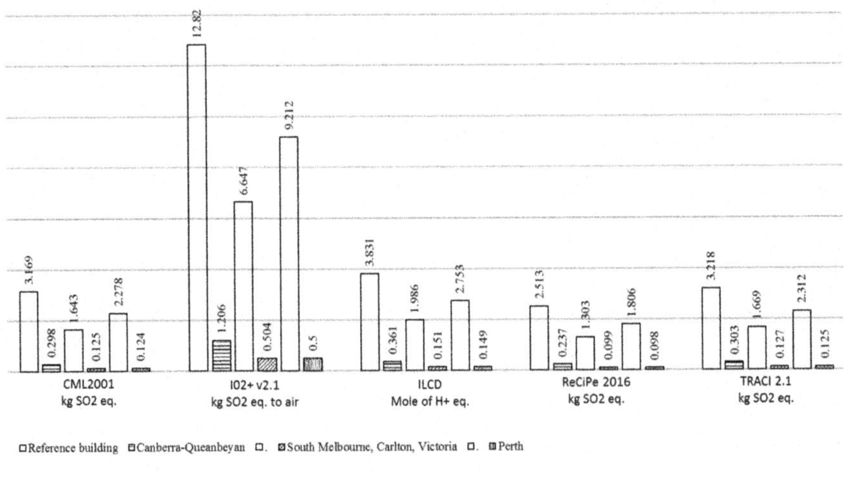

Figure 5.7 Acidification potential impact

major acid substances in comparison with the other structures, at values of 0.323 and 2.299 kg SO2-eq per square meter, respectively. However, with ILCD and TRACI analysis, the higher results stem from the structure with the steel roof frame with the value of 0.502 mole of H^+ eq (ILCD method) and 0.442 kg SO2-eq per square meter (TRACI method). Along with reduced energy consumption in comparison with the reference building, the environmental impact also is decreased with the ratio that helps these case studies achieve maximum points in Credit 15 elements.

Figure 5.8 illustrates the combined results of environmental impact, which is calculated for the energy consumption and materials using the case studies. These results also show that energy consumption is the critical aspect that should be considered to reduce the environmental impact during the life-cycle of the buildings in the case studies. Timber material contributes the minimum effect during the building lifetime in all the impact because the woodwork quantity in these case studies is not significant, and the emission embodied of timber is smallest among the studied materials (Morel et al., 2001; Taylor & Van Langenberg, 2003). In the case studies, there are 95% of timber products that meet the stipulated formaldehyde level and stem from a reused source, which leads to the achievement of 2 credit points in Green Star Credit 13 and 20 (Green Building Council of Australia, 2015a).

Figure 5.8 also illustrates that the environmental impact is derived mostly from energy consumption. However, there is some distinction; for example, ozone depletion potential results are dissimilar among the used methods. Substance impact results are quite similar for CML2001, I02+, ILCD, and TRACI categories when the highest impact result belongs to steel in all three case studies. Meanwhile, analyses by the ReCiPe method are significantly different; nevertheless, the highest impact is still derived from energy consumption. The reduced use of steel reinforcement in the buildings helped the projects achieve two credit points.

The three case studies conducted life-cycle assessment of the projects and the reference building to compare the environmental impact reduction against all significant environmental impact categories mentioned in Green Star Credit 19A (Figure 5.8). These analyses helped these projects achieve an additional seven points to the accumulated credit points.

Table 5.8 Model's output validation against case studies' GWP impacts (kg CO2 eq/m^2)

Method	*Case 1*	*Case 2*	*Case 3*
Case studies' GWP impact results	290	380	440
CML2001 – Jan. 2016, Global Warming Potential (GWP 100 years)	297.64	381.56	447.11
I02+ v2.1 – Global warming 500 yrs. – Midpoint	289.20	370.32	433.71
ILCD – Climate change midpoint, incl. biogenic carbon (v1.09)	296.65	380.26	445.52
ReCiPe 2016 Midpoint (H) – Climate change, incl. biogenic carbon	288.09	369.00	432.03
TRACI 2.1, Global Warming Air, incl. biogenic carbon	296.56	380.24	445.37

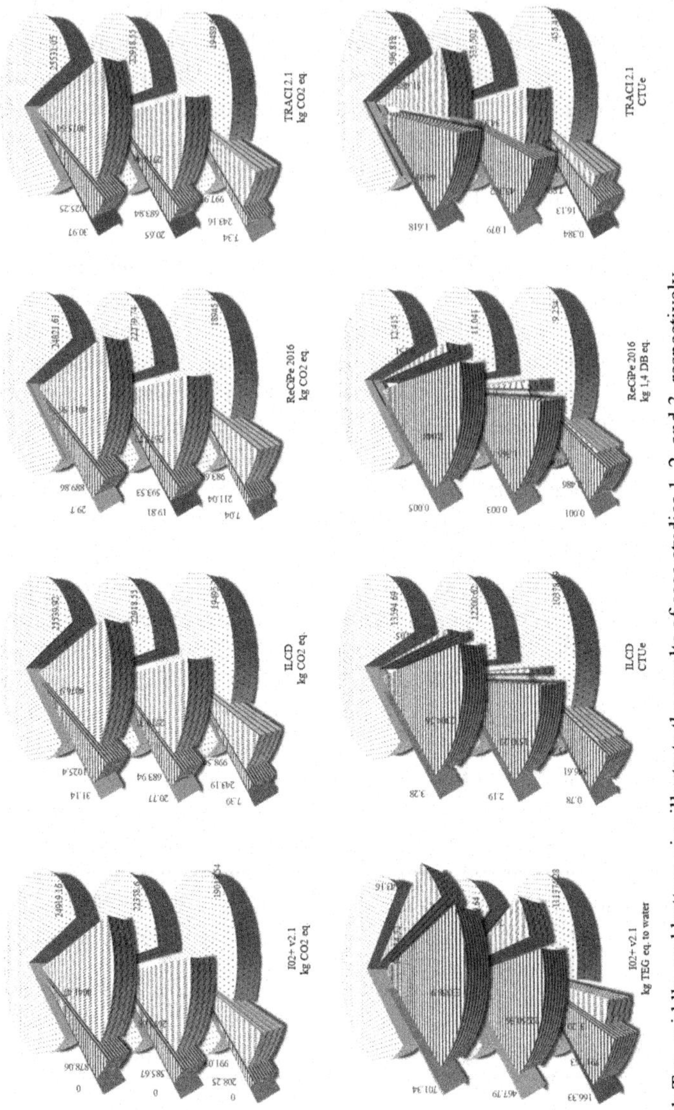

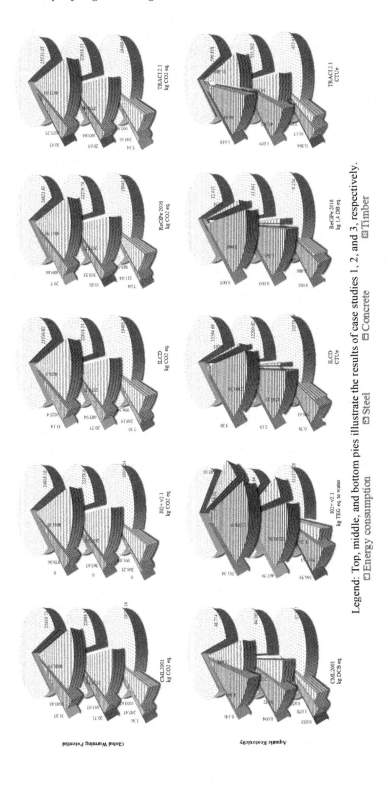

Legend: Top, middle, and bottom pies illustrate the results of case studies 1, 2, and 3, respectively.
Energy consumption   Steel   Concrete   Timber

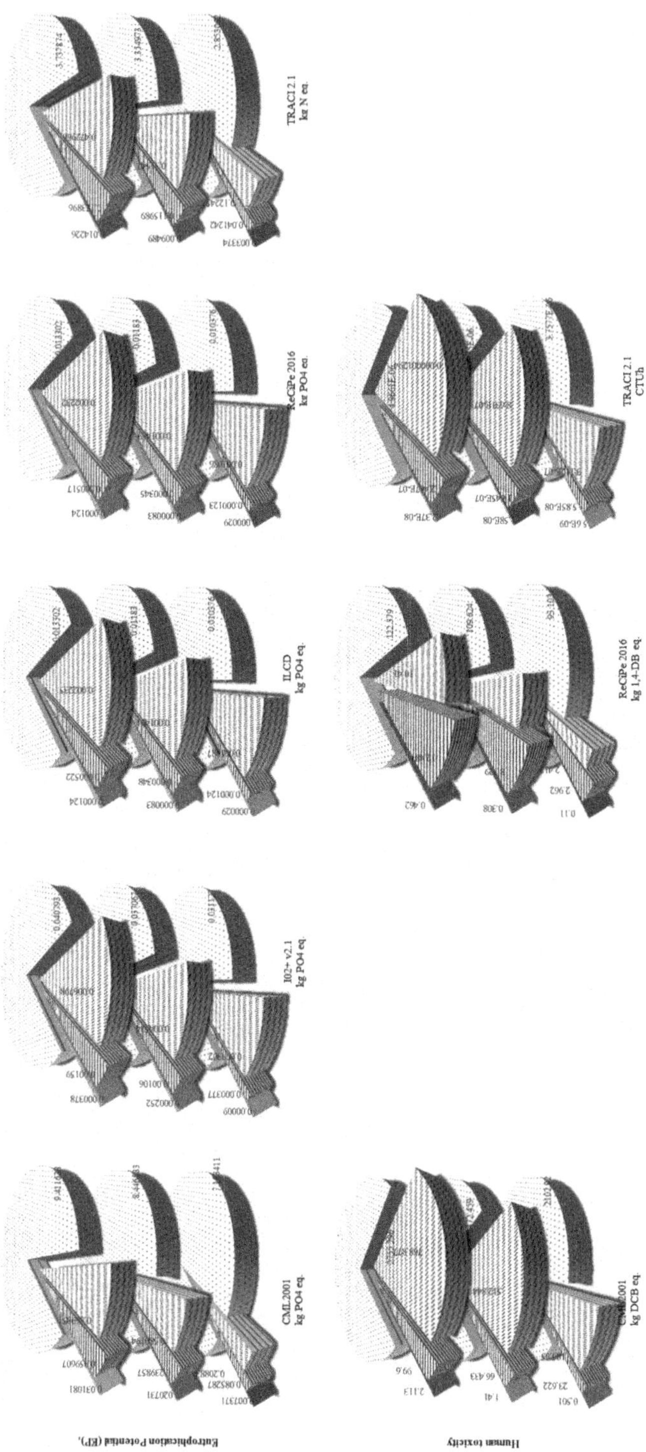

Legend: Top, middle, and bottom pies illustrate the results of case studies 1, 2, and 3, respectively.
Energy consumption Steel Concrete Timber

Figure 5.8 (Continued)

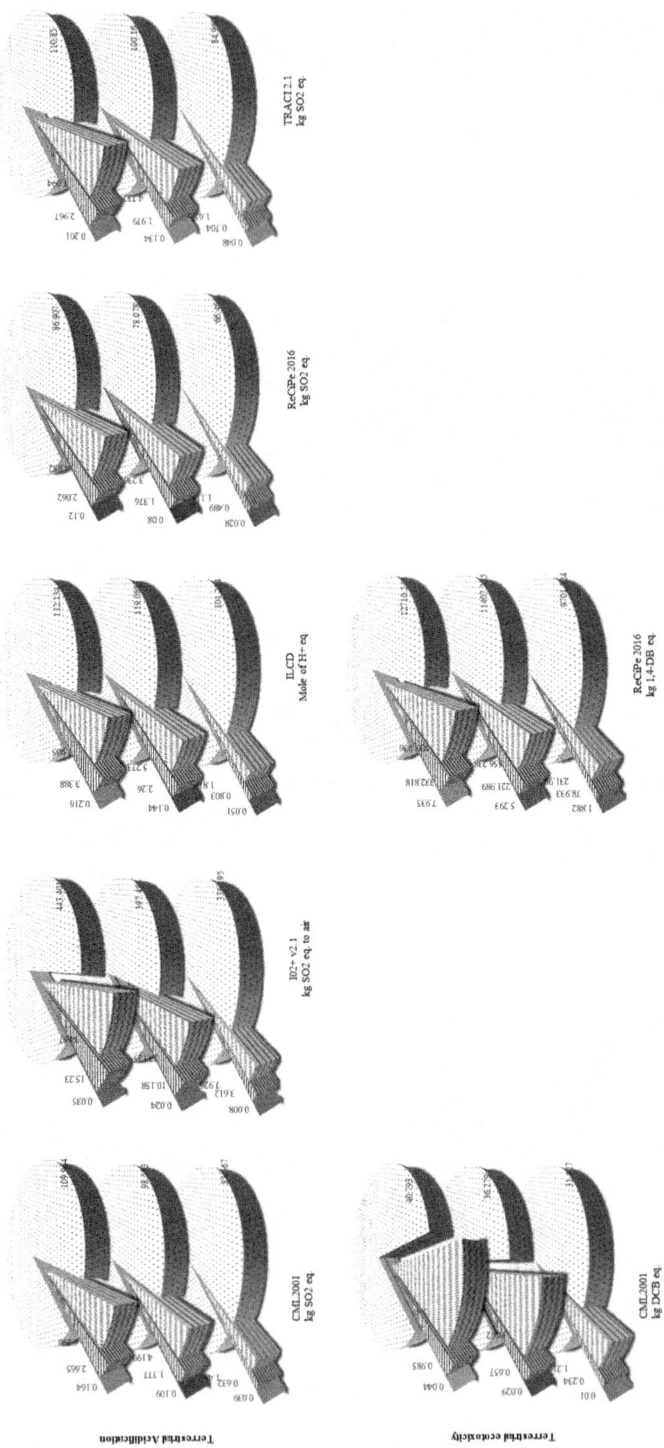

Legend: Top, middle, and bottom pies illustrate the results of case studies 1, 2, and 3, respectively.
Energy consumption   Steel   Concrete   Timber

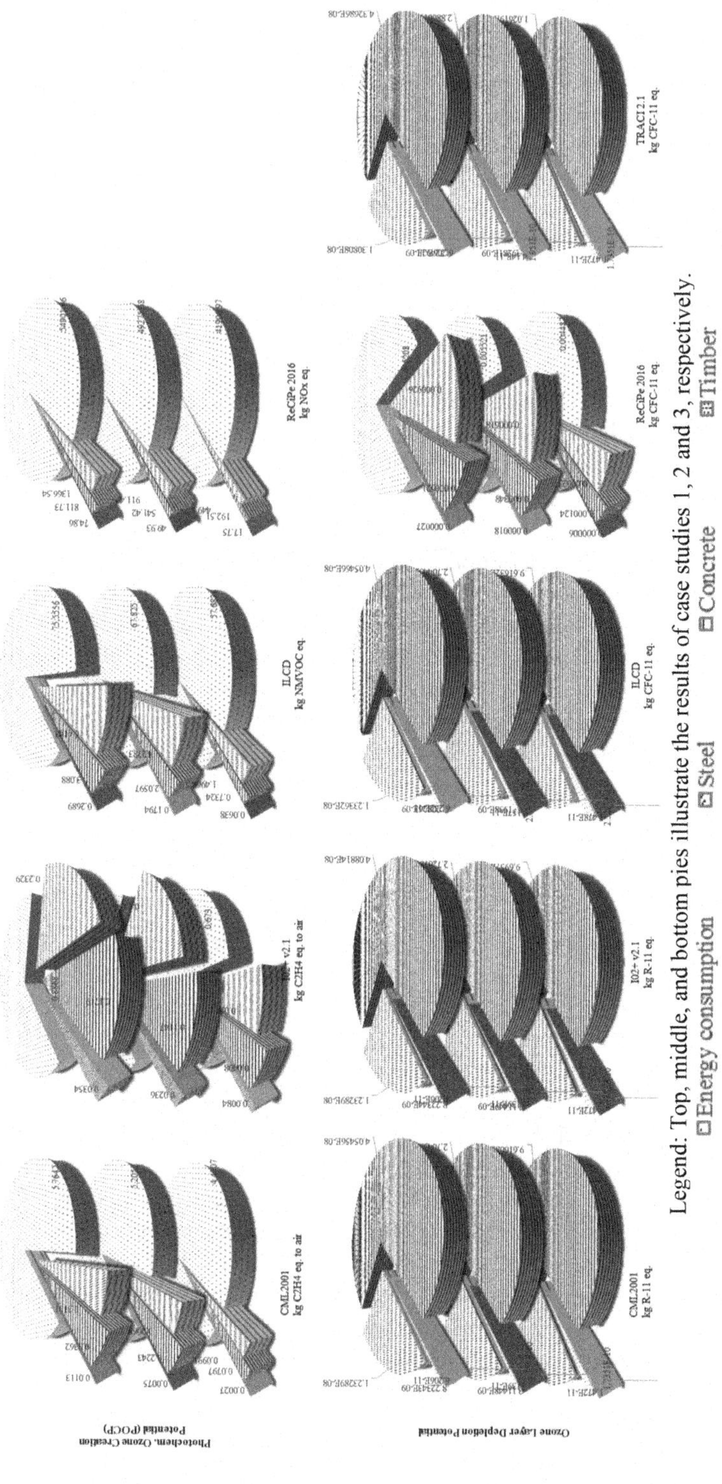

Figure 5.8 Case studies on environmental impact results

The ratio of the concrete mixture in the case study 1 is 60% cement and 40% silica fume replacement. The other two case studies use fly ash as supplementary cementitious materials with similar cement content. This ratio of cement and replaced materials helps the projects achieve 2 points in Green Star Credit 19B, which is related to the reduction of Portland cement content. The projects also achieved 1 credit point because crushed slag aggregate was used in the case study 1, and manufactured sand was used for the case studies 2 and 3, along with the 50% water reduction in the concrete mixture.

The proposed model helps designers achieve life-cycle greenhouse gas emissions assessments along with the achieved credit points by the estimation. The results demonstrate that the greenhouse gas emissions of the buildings are less than those of the equivalent reference building. The model's output, which is presented in Table 5.8, is fairly similar to the case studies' results, which are illustrated in Table 5.1.

Apart from the GWP impact comparison, Figure 5.8 shows the ability of the model to extract other environmental impact data during the lifetime of the building, as well as Green Star credit point calculation for the credits related to greenhouse gas emission reduction.

This determination helps the projects achieve 34 credits points with respect to energy consumption reduction as well as greenhouse gas emission reduction in the Green Star rating tool, which was mentioned in Section 3.4. It emphasizes that the model is constantly verified using different datasets and material under different conditions to minimize its computation errors. This process is understandably lengthy and quite intensive. Within the scope of this book, the verification for greenhouse gas emission analysis has been successfully achieved. By changing the building envelopes, concrete mixtures, and the ratio of steel and timber, the book will help designers understand and achieve credit points in Green Star.

5.4 Summary

In this chapter, the validation of the life-cycle greenhouse gas emissions and achievement of credit points used by the model of this book is discussed. The results of the case studies analyzed by the model show the ability to conduct the requirements of the criteria of greenhouse gas reduction during the life of the buildings, which contribute 34 credit points of the total assessed credit points for the projects.

In addition, the model shows the ability to provide the optimal option considering the other sources of environmental impacts, which could be requested to assess. The results of case studies are compared to the reference building. The obtained values from the model are cross-checked and lead to the accurate and consistent validation of greenhouse gas emission analysis by the model.

Thus, this proposed life-cycle greenhouse gas emissions model offers resolutions to planners and designers deliberating optimal options in terms of lower life-cycle greenhouse gas emissions and higher Green Star credit point achievement.

6 Conclusion

6.1 Introduction

The book aimed to develop a computational model to investigate life-cycle greenhouse gas emissions of a commercial building in Australia in terms of its ability to achieve Green Star certification.

This chapter divides into four parts. In the first part, the significant conclusions of the book are discussed, followed by relevant recommendations derived from these conclusions. The research also had some specific limitations, which are described in this chapter. Finally, future study directions that could evolve from this research are presented in detail.

6.2 Major conclusions

The contribution of green building design to sustainable development is undeniable. This is one of the many practical activities in the construction field that are aimed at reducing the environmental impacts. The approach presented here will help reduce greenhouse gas emissions from many significant sources during the project stages of production, construction, operation, and other end-of-life stages of the building. The green building concept solves considerable concerns regarding sustainable economic and environmental development, and healthy habitat in the building.

Chapter 2 of this book discussed sustainable trends in the construction industry worldwide and in Australia, as well as the procedure of life-cycle greenhouse gas emissions assessment in green building projects. The Green Building Council of Australia recognizes the design of a green building as the measure that reduces the building's environmental impact. According to this definition, green buildings promote the building efficiency to enhance better performance, reduce the amount of life-cycle greenhouse gas emissions significantly, and raise awareness about reducing the energy and materials used in the building. The life-cycle assessment adopts the cradle-to-grave method to assess the environmental impacts of a product. Greenhouse gas emissions are considered as the project output to the building outdoor environment (Ramesh et al., 2010; Chau et al., 2015). In order to efficiently improve a greenhouse gas emissions simulation model for a complex commercial building, life-cycle assessment should be investigated with straightforward and interactive programs after or in the process of construction design. The combination of design has followed the requirements defined in "green" standards, for example, credit points achievement of Green Star (Design & As Built) in Australia, and the building's life-cycle assessment including life-cycle greenhouse gas emissions, life-cycle energy consumption, and life-cycle cost.

The literature discussed the development of the model techniques to assess life-cycle greenhouse gas emissions. Hundreds of programs have been developed that contain many methods for estimating the environmental impacts in the life-cycle of construction projects (Figure 2.5). Tam et al. (2018a) indicated that GaBi and SimaPro are the two most outstanding programs that provide alternatives to estimate not merely energy consumption, but also the environmental impacts of energy-related elements (building fabric, glazing, air conditioning and ventilation systems, heat water supply, etc.) during the green building lifetime. In this research, the assessment model was developed in GaBi version 8.7 to perform all essential criteria involving air pollutant reduction in a green building in Australia.

In Chapter 3, the book analyzed all the credits in the Green Star (Design & As Built) version 1.1 rating tool, to understand which credits are directly involved with the process to reduce life-cycle greenhouse gas emissions of commercial buildings in Australia. During the lifetime of a typical building, material production and machinery and equipment operations generate many types of air pollutant emissions. The application of the Green Star rating tool was compatible with the analysis of the building's element processes. The model helped the green building projects achieve the highest score of 34 credit points of all the assessed credit points. The impacts of energy consumption and the employment of major materials (concrete, steel, and timber) in the projects, which are correlated with credit 13 – Indoor pollutant, credit 15 – Greenhouse gas emissions, credit 19 – Life-cycle impacts, and credit 20 – Responsible Building Material in the Green Star rating tool, are also assessed by the model.

The analysis of life-cycle greenhouse gas emissions was carried out based on the element unit estimation, which includes the processes from manufacture, construction, operation, and other end-of-life stages of the building. Then, the model used the quantity breakdown from the building design. The final results were calculated by Equation 3.1 and Equation 3.2 presented in Chapter 3. As a result of the application and satisfying the Green Star rating tool's requirements, the final step of the model was awarding the points for each related credit to the green building project.

Chapter 3 provided a range of data of materials and building envelopes such that designers and practitioners can choose the optimal option for their design that matches with many typical climate zones in Australia. In this research, seven climate zones in Australia – Adelaide, Brisbane, Canberra, Perth, Sydney, Hobart, and Melbourne – were selected for analysis. The data of the building envelope's thermal resistance are collected from the handbook of the Insulation Council of Australia and New Zealand (2014). This handbook presents 13 roof types, 11 wall types, and 3 floor types, which contain the main structure with main materials such as steel, concrete, and timber, as well as other thermal-resisting materials (Table 3.10).

The research considered all concrete types with the typical compressive strengths in the range of 20–100 MPa, as well as the appropriate proportions of supplementary cement materials, alternative aggregates, and the amount of water used in concrete (Green Building Council of Australia, 2015a). The model helped designers obtain the concrete mixture design following the desired concrete strength for their project designs. Because of massive importing and exporting data, the model collected the data in Microsoft Excel and then stimulated the calculation processes in Visual Basic. From there, the project can achieve the maximum available credit points of the Green Star rating tool.

The validation of the case studies presented in Chapter 5 showed the ability of the model to adjust geographically relevant temperatures following the Australian Bureau of Meteorology data. The three case studies, which are in three different locations in Australia, were compared in Chapter 5 to evaluate the environmental impacts of the materials and building envelope alternatives. The life-cycle greenhouse gas emissions of the studied parameters in terms of Green Star credits were estimated and validated, proving the applicability and flexibility of the model to adjust to different zones and locations in Australia.

The advantages of this model are the adaptation ability within the Australian construction context. The model was developed to follow the regulations outlined in the National Construction Code of Australia (NCC) under different climate zones. The energy consumption was calculated and summarized for typical building envelopes as specified under the Insulation Council of Australia and New Zealand (2014) handbook. This proposed model, built under GaBi software version 8.0, contains four conventional methodologies: ReCiPe, ILCD, TRACI, and Impact 2002+. According to the literature, the selection of material significantly affects a building's energy consumption. Life-cycle greenhouse gas emissions will thus change if the projects use materials with different temperature resistance values.

The other benefit of this model is that designers can determine the greenhouse gas emission levels produced during a project's life according to the required energy consumption levels. This research addressed the Green Star requirements, which are broadly regulated in Australia. The model also contributes to the Australian construction industry with the ability to adapt to the Green Star energy and greenhouse gas emissions reduction scenarios during the building's life-cycle. The model developed in this book was aimed at helping designers compare their construction project with the reference building. From there, they can estimate the elements' credit points that relate to energy consumption and greenhouse gas emissions reduction for their project in comparison with the reference building.

This proposed assessment procedure also helped them achieve reward points for their project under the "Life-cycle impacts" credit, because the project life-cycle impacts assessment is carried out according to NCC requirements. The additional amount of greenhouse gas emissions and energy consumption deduction can thus pave the way for green buildings to play a significant role in sustainable development. Moreover, the model provided the project's stakeholders with a shortcut to assess eight environmental impact categories mentioned in Chapter 3, which meet the requirements in the Green Star "Life-cycle impacts" criterion. This criterion let the project achieve another six points. Hence, the maximum credit points from this book are 34 out of 100 points of the rating tool. Using this model, life-cycle assessment practitioners can compare and select the optimal sustainable impact design and hence achieve Green Star certificates for their projects.

6.3 Recommendations

This book aimed to propose a supporting tool for building envelope options, particular concrete types, to achieve Green Star credit points. Concrete design using this model was dependent on practical applications and conditions of a concrete batching plant. Upon adjustment, greenhouse gas emissions can be readily re-estimated using the model. In this book, even though the use of supplementary cementitious materials will lessen the burden on the environment, alternative materials also present different perspectives. For example, while silica fume is the optimal material to reduce greenhouse gas emissions, it costs more than the other two supplementary cementitious materials. That means the demand for cement replacement will increase if the market price of these materials is consistent with the project's budget plan and the environmental protection objectives of the investor. As such, extra research efforts are required to thoroughly examine specific materials under Green Star credits regarding greenhouse gas emissions reduction.

It is clear that the proposed model can be effectively and efficiently employed to calculate life-cycle greenhouse gas emissions. The proposed model can be further extended to compute life-cycle greenhouse gas emissions for other Green Star credits, such that a systematic and automated approach to achieving Green Star status can be established. The proposed model thus presents a reliable application to automate Green Star credit achievement for the future.

The research recommended that the structures that combine timber with other materials have less severe environmental impacts than those using metal, brick, or concrete. Life-cycle greenhouse gas emissions from structures using the combination of concrete and brick are higher than the ones using only concrete (Le et al., 2018b). The book's results were similar to existing results in the literature demonstrating that bricks have higher negative environmental impact than concrete products. Using structures with higher total R-value is one of the requirements for green buildings, which will reduce their energy consumption. Consequently, life-cycle greenhouse gas emissions will be decreased along with the reduction of greenhouse gas emissions, generated during production and operational phases.

This research was conducted in the Australian context. Energy consumption is used for calibrating indoor environment at a given outdoor temperature in different climate zones. All energy consumption possibilities have been illustrated in the literature. Energy scenarios during the operation phase, followed by the requirement of the "Life-cycle impacts" and "Energy" credits in Green Star, play a significant role in life-cycle greenhouse gas emissions assessment.

This book has focused on the analyses and requirements under Green Star. Environmental impacts of the life-cycle of materials and energy consumption reduction are assessed for typical building fabrics. The chapter provided insights into sustainable construction and assistance to practitioners in recognizing the environmental contribution of the materials involved in building structures during a project's lifetime. This book has also revealed the effects of combining building envelopes with greenhouse gas emissions, and energy consumption reduction under strict Green Star regulation. From the book, it can be suggested that construction projects can be granted maximum credit points under Green Star credits regarding the reduction of greenhouse gases. The additional amount of greenhouse gas deduction and reduction of energy consumption can thus pave the way for green buildings to play a significant role in sustainable development.

6.4 Limitations of the research

Albeit the book's effectiveness and efficiency, there are some limitations to this research, which are as follows:

(1) After collecting the data from GaBi, the model only ran under the environment of MS Excel to ensure valid syntax for MS Visual Basic commands.
(2) The model was developed to assess the impacts of building envelope alternatives on the building's energy consumption, as well as greenhouse gas emissions reduction. The information on the building envelope alternatives was collected from the handbook of Insulation Council of Australia and New Zealand (2014).
(3) Although the model could be employed to design the concrete mixture at the factory or on site, it still needs to be adjusted to match functional constituent materials such as cement, aggregate types, admixtures, and water content. This means that initial conditions play vital roles in obtaining the model's accuracy. In fact, the model's weakness is its dependence on extra initial conditions, which are driven by a specific credit. According to the research findings, the inclusion of new initial conditions can cause internal and challenging conflicts that may require optimal resolution.
(4) The proposed greenhouse gas emissions were calculated for concrete with strengths in the range of 20–100 MPa. For the strengths of concrete outside this range, variables and parameters in the model also need to be updated. A similar situation occurs when practitioners wish to employ supplementary cementitious materials other than the three supplementary cementitious materials proposed in Green Star and studied in this research. Also, even though greenhouse gas emissions were calculated from the concrete's

life-cycle, following Green Star requirements, concrete masonry is excluded from this book.

(5) The model only calculated greenhouse gas emissions during the lifetime of a building. The other life-cycle parameters such as life-cycle cost and life-cycle energy consumption will be studied in a separate study.

(6) The model's accuracy was clearly dependent on specific applications, and the resulting sensitivity is thus dependent on initial conditions. Because this research was devoted to computing life-cycle greenhouse gas emissions for Green Star only, further detailed sensitivity analyses will be presented in a separate study.

6.5 Future study directions

To achieve sustainable goals, the analysis of life-cycle greenhouse gas emissions needs to be taken into consideration along with the life-cycle cost assessment of green buildings. The future research will collect data on other factors of building's components in its lifetime and the Australian green building context. Databases for both life-cycle greenhouse gas emissions and cost of commercial buildings will be developed to achieve that purpose.

Also, besides Green Star – Design & As Built, there is also a variety of green building environmental rating systems worldwide in general, and in Australia such as "the Green Star – Interiors", "the Green Star – Communities", and "the Green Star – Performance". In the next step of the research development, this book will be implemented and adapted further to satisfy the requirements of other sustainable tools worldwide in future studies.

Although the four methodologies, namely, ReCiPe 2016, TRACI 2.1, ILCD, and Impact 2002+, are the updated methods to analyze environmental impacts, other methodologies will be considered for analyses in further research.

Certain structure types such as residential houses and industrial projects are rarely assessed with the green building approach, due to some barriers that prevent these projects from being granted the green building certificate. Therefore, further studies will identify these obstacles and recommend available resolutions to enhance sustainable development in the construction industry.

References

Abbas, A., Fathifazl, G., Isgor, O. B., Razaqpur, A. G., Fournier, B. & Foo, S. 2006, 'Environmental benefits of green concrete', in *2006 IEEE EIC Climate Change Conference*, 10–12 May 2006, pp. 1–8.

Abd Rashid, Ahmad Faiz & Yusoff, Sumiani 2015, 'A review of life-cycle assessment method for building industry', *Renewable and Sustainable Energy Reviews*, vol. 45, pp. 244–8.

Acero, Aitor P., Rodríguez, Cristina & Ciroth, Andreas 2017, *LCIA methods: Impact assessment methods in life-cycle assessment and their impact categories*, GreenDelta GmbH, Berlin, vol. 23.

ACI Committee 2008, *ACI 211.4R-08: Guide for selecting proportions for high-strength concrete using Portland cement and other cementitious materials*, American Concrete Institute, Farmington Hills, MI.

Adibi, N., Lafhaj, Z., Yehya, M. & Payet, J. 2017, 'Global resource indicator for life-cycle impact assessment: Applied in wind turbine case study', *Journal of Cleaner Production*, vol. 165, no. Supplement C, pp. 1517–28.

Ahn, Yong Han, Pearce, Annie R., Wang, Yuhong & Wang, George 2013, 'Drivers and barriers of sustainable design and construction: The perception of green building experience', *International Journal of Sustainable Building Technology and Urban Development*, vol. 4, no. 1, pp. 35–45.

Aïtcin, Pierre-Claude & Mindess, Sidney 2011, *Sustainability of concrete*, Spon Press, New York, NY.

Al-Ghamdi, Sami G. & Bilec, Melissa M. 2016, 'Green building rating systems and whole-building life-cycle assessment: Comparative study of the existing assessment tools', *Journal of Architectural Engineering*, vol. 23, no. 1, p. 04016015.

Allen, Joseph G., Bernstein, Ari, Eitland, Erika, Cedeno-laurent, Jose, MacNaughton, Piers, Spengler, John D. & Augusta, Williams 2017, *The nexus of green buildings, public health and the UN sustainable development goals*, Harvard T.H. Chan School of Public Health, Boston, MA.

Álvarez-Herránz, Agustín, Balsalobre, Daniel, Cantos, José María & Shahbaz, Muhammad 2017, 'Energy innovations-GHG emissions nexus: Fresh empirical evidence from OECD countries', *Energy Policy*, vol. 101, pp. 90–100.

Alyami, Saleh H. & Rezgui, Yacine 2012, 'Sustainable building assessment tool development approach', *Sustainable Cities and Society*, vol. 5, pp. 52–62.

Amara, Fatima, Agbossou, Kodjo, Cardenas, Alben, Dubé, Yves & Kelouwani, Sousso 2015, 'Comparison and simulation of building thermal models for effective energy management', *Smart Grid and Renewable Energy*, vol. 6, no. 4, p. 95.

American Concrete Institute 2014, *ACI 318–14 building code requirements for structural concrete and commentary*, American Concrete Institute, Farmington Hills, MI.

American Society for Testing and Materials 2013, *C33/C33M-13: Standard specification for concrete aggregates*, American Society for Testing and Materials, Philadelphia, PA.

American Society of Heating Refrigerating and Air-Conditioning Engineers 2006, 'Green/sustainable high-performance design', in ASHRAE (ed.), *The ASHRAE green guide*, 2nd edn, Butterworth-Heinemann, Burlington, pp. 3–16.

American Society of Heating Refrigerating and Air-Conditioning Engineers Press 2006, *ASHRAE green guide: The design, construction, and operation of sustainable buildings*, Butterworth-Heinemann, Oxford, United Kingdom.

Andersson, Göran 2013, *Environmental evaluation of steel and steel structures*, Jernkontoret, Stockholm.

Architecture & Design 2016, *10 software tools for green design*, SimaPro, viewed 30.10.2017, <www.architectureanddesign.com.au/news/bpn/environ/10-software-tools-for-green-design>.

Asdrubali, Francesco, Baldassarri, Catia & Fthenakis, Vasilis 2013, 'Life-cycle analysis in the construction sector: Guiding the optimization of conventional Italian buildings', *Energy and Buildings*, vol. 64, pp. 73–89.

Asif, M., Muneer, T. & Kelley, R. 2007, 'Life-cycle assessment: A case study of a dwelling home in Scotland', *Building and Environment*, vol. 42, no. 3, pp. 1391–4.

Australia Government 2012, *Construction and demolition waste guide – Recycling and reuse across the supply chain*, Australian Government, Australia, viewed <www.environment.gov.au/system/files/resources/b0ac5ce4-4253-4d2b-b001-0becf84b52b8/files/case-studies.pdf>.

Australia Government 2016, *Australia's emissions projections 2016*, Commonwealth of Australia, Canberra, Australia.

Australian (Iron & Steel) Slag Association 2012, *Blast furnace slag cements & aggregates: Enhancing sustainability*, Australasian (Iron & Steel) Slag Association, NSW, Australia.

Australian Building Codes Board 2006, *ABCB protocol for building energy analysis software*, Australia, viewed 30.03.2018, <www.abcb.gov.au/Resources/Publications/Education-Training/ABCB-Protocol-for-Building-Energy-Analysis-Software>.

Australian Building Codes Board 2013, *National construction code of Australia – Class 1 and 10 buildings – Volume two, Australian Building Codes Board*, viewed <www.gbca.org.au/uploads/195/3267/Green_Star_-_Greenhouse_Gas_Emissions_Calculator_Guide.pdf>.

Australian Building Codes Board 2016, *NCC 2016 building code of Australia*, Australia, viewed.

Australian Bureau of Statistics 2013, *Measures of Australia's progress*, Australian Bureau of Statistics, viewed 12.01.2017, <www.abs.gov.au/ausstats/abs@.nsf/mf/1370.0>.

Australian Government 2008, *National pollutant inventory, emission estimation technique manual for combustion engines*, Version 3.0, Department of the Environment, Water, Heritage and the Arts, Australian Government, Canberra, Australia.

Australian Institute of Quantity Surveyors & Master Builders Australia 2016, *Australian standard method of measurement of building works*, Australian Institute of Quantity Surveyors, Sydney, NSW, Australia.

Australian Inventory data project 2009, *Australian unit process life-cycle inventory*, Life-cycle Strategies Pty. Ltd, Australia, viewed 01.08.2017, <www.auslci.com.au/index.php/datasets/Materials>.

Australian Life-cycle Assessment Society (ALCAS) 2017, *The Australian National Life-cycle Inventory Database (AusLCI)*, Australian Life-cycle Assessment Society (ALCAS), viewed 30.10.2017, <www.auslci.com.au/index.php/Datasets>.

Autodesk Sustainability Workshop 2017, *Whole building energy analysis*, Autodesk, viewed 30.10.2017, <https://sustainabilityworkshop.autodesk.com/buildings/whole-building-energy-analysis>.

Azapagic, Adisa 2006, 'Life-cycle assessment as an environmental sustainability tool', *Renewables-Based Technology: Sustainability Assessment*, pp. 87–110.

Bare, Jane C. 2011, 'TRACI 2.0: The tool for the reduction and assessment of chemical and other environmental impacts 2.0', *Clean Technologies and Environmental Policy*, vol. 13, no. 5, pp. 687–96.

Bare, Jane C. 2012, *Tool for the reduction and assessment of chemical and other environmental impacts TRACI 2.1: User's manual*. United States Environmental Protection Agency, SOP No. S-10637-OP-1-0.

Bare, Jane C., Pennington, D. W. & Udo de Haes, H. A. 2000, *An international workshop on life-cycle impact assessment sophistication*, EPA/600/R-00/023.

Bare, Jane C., Young, Daniel, Q. A. M. S., Hopton, Matthew & Chief, S. A. B. 2012, *Tool for the Reduction and Assessment of Chemical and other Environmental Impacts (TRACI)*, US Environmental Protection Agency, Washington, DC.

Basbagill, J., Flager, F., Iepech, M. & Fischer, M. 2013, 'Application of life-cycle assessment to early-stage building design for reduced embodied environmental impacts', *Building and Environment*, vol. 60, pp. 81–92.

Bickley, John A., Hooton, R. Doug & Hover, Kenneth C. 2006, 'Performance specifications for durable concrete', *Concrete International*, vol. 28, no. 9, pp. 51–7.

Biermann, Frank, Kanie, Norichika & Kim, Rakhyun E. 2017, 'Global governance by goal-setting: The novel approach of the UN sustainable development goals', *Current Opinion in Environmental Sustainability*, vol. 26–27, pp. 26–31.

Bird, Tom 2018, *World steel recycling in figures 2013–2017*, Bureau of International Recycling, Brussels.

Blom, Inge, Itard, Iaure & Meijer, Arjen 2011, 'Environmental impact of building-related and user-related energy consumption in dwellings', *Building and Environment*, vol. 46, no. 8, pp. 1657–69.

Blue Environment Pty Ltd 2014, *Tasmanian waste review – Final report: P403*, Prepared by Blue Environment Pty Ltd, for Waste Advisory Committee, Tasmania.

Bon-Gang, Hwang & See, Tan Jac 2012, 'Green building project management: Obstacles and solutions for sustainable development', *Sustainable Development*, vol. 20, no. 5, pp. 335–49.

Brause, Rüdiger 2010, *Adaptive Modellierung und Simulation*, Frankfurt.

Breins Engineering 2016, *Coarse aggregate*, viewed 24.08.2017, <www.buildingresearch.com.np/services/ct/ct4.php>.

Briassoulis, Demetres, Dejean, Cyril & Picuno, Pietro 2010, 'Critical review of norms and standards for biodegradable agricultural plastics, part II: Composting', *Journal of Polymers and the Environment*, vol. 18, no. 3, pp. 364–83.

Briassoulis, Helen 2001, 'Sustainable development and its indicators: Through a (Planner's) Glass Darkly', *Journal of Environmental Planning and Management*, vol. 44, no. 3, pp. 409–27.

British Standards Institution 2011, *PAS 2050:2011 specification for the assessment of the life-cycle greenhouse-gas emissions of goods and services*, Department for Environment, Food and Rural Affairs, & British Standards Institution, United Kingdom, viewed.

Broun, Reza & Menzies, Gillian F. 2011, 'Life-cycle energy and environmental analysis of partition wall systems in the UK', *Procedia Engineering*, vol. 21, pp. 864–73.

Brulliard, Christophe, Cain, Rebecca, Do, Daphne, Dornom, Tim, Evans, Katherine, Lim, Brendan, Olesson, Erica & Young, Suzi 2012, *The Australian recycling sector*, Department

of Sustainability, Environment, Water, Population and Communities (DSEWPaC), Australia.
Brundtland, Gro 1987, *Our common future: Report of the 1987 World Commission on Environment and Development*, United Nations, Oslo, vol. 1, p. 59.
Buchanan, Andrew H. & Levine, S. Bry 1999, 'Wood-based building materials and atmospheric carbon emissions', *Environmental Science & Policy*, vol. 2, no. 6, pp. 427–37.
Buckley, B. & Logan, K. 2016, *World green building trends 2016: Developing markets accelerate global green growth*, Dodge Data & Analytics, Bedford, MA.
Bueno, Cristiane, Hauschild, Michael Zwicky, Rossignolo, João Adriano, Ometto, Aldo Roberto & Mendes, Natália Crespo 2016, 'Sensitivity analysis of the use of life-cycle impact assessment methods: A case study on building materials', *Journal of Cleaner Production*, vol. 112, pp. 2208–20.
Building Research Establishment Environment Assessment Method 2015, *What is BREEAM?*, Building Research Establishment (BRE), viewed 31.08.2015, <www.breeam.org/>.
Bunning, Jessica, Beattie, Colin, Rauland, Vanessa & Newman, Peter 2013, 'Low-carbon sustainable precincts: An Australian perspective', *Sustainability*, vol. 5, no. 6, pp. 2305–26.
Burgess, A. A. & Brennan, D. J. 2001, 'Application of life-cycle assessment to chemical processes', *Chemical Engineering Science*, vol. 56, no. 8, pp. 2589–604.
Butcher, Ken J. 2015, *CIBSE guide A: Environmental design*, 7th edn, CIBSE, <http://app.knovel.com/hotlink/toc/id:kpCIBSEGA8/cibse-guide-environmental/cibse-guide-environmental>.
Cabeza, Luisa F., Rincón, Lídia, Vilariño, Virginia, Pérez, Gabriel & Castell, Albert 2014, 'Life-cycle assessment (LCA) and Life-cycle energy analysis (LCEA) of buildings and the building sector: A review', *Renewable and Sustainable Energy Reviews*, vol. 29, pp. 394–416.
Caldarone, Michael A. 2009, *High-strength concrete: A practical guide*, 1st edn, Taylor & Francis, London.
Carre, A. 2011, *A comparative life-cycle assessment of alternative constructions of a typical Australian house design*, Forest and Wood Products Australia, Project Number PNA 147–0809.
Cassidy, R., Wright, G. & Flynn, L. 2003, *White paper on sustainability: A report of the green building movement. building design and construction*, Reed Business Information. Clearwater, <www.usgbc.org/Docs/Resources/BDCWhitePaperR2.pdf>.
Cement Concrete & Aggregates Australia 2007, *Manufactured sand*, Cement Concrete & Aggregates Australia, Australia, viewed 30.10.18, <https://www.ccaa.com.au/iMIS_Prod/CCAA/Public_Content/PUBLICATIONS/Reports/Manufactured_Sand_National_Test_Methods_And_Specification_Values.aspx>.
Cement Concrete & Aggregates Australia 2008, *Use of recycled aggregates in construction*, Cement Concrete & Aggregates Australia, viewed 30.10.18, <https://www.ccaa.com.au/iMIS_Prod/CCAA/Public_Content/PUBLICATIONS/Reports/Use_of_Recycled_Aggregates_In_Construction.aspx>.
Cement Concrete & Aggregates Australia 2010, *Sustainable concrete materials*, Cement Concrete & Aggregates Australia, Australia, viewed 30.10.18, <https://www.ccaa.com.au/iMIS_Prod/CCAA/Public_Content/PUBLICATIONS/Technical_Publications/Briefings/Briefing_11_-_Sustainable_Concrete__Materials.aspx>.
Chau, C. K., Leung, T. M. & Ng, W. Y. 2015, 'A review on life-cycle assessment, life-cycle energy assessment and life-cycle carbon emissions assessment on buildings', *Applied Energy*, vol. 143, no. Supplement C, pp. 395–413.

Cheung, Sai On 2013, 'Special issue on green and sustainable construction projects: The facets of sustainability', *Journal of Legal Affairs and Dispute Resolution in Engineering and Construction*, vol. 5, no. 4, pp. 162–162.

Cho, Su-Hyun & Chae, Chang-U. 2016, 'A study on life-cycle CO2 emissions of low-carbon building in South Korea', *Sustainability*, vol. 8, no. 6, p. 579.

Chowdhury, Ashfaque Ahmed, Rasul, M. G. & Khan, M. M. K. 2007, 'Modelling and simulation of building energy consumption: A case study on an institutional building in central Queensland, Australia', in *Proceedings of 10th International Building Performance Simulation Association (IBPSA) International Conference*, pp. 3–6.

Collins, Frank 2010, 'Inclusion of carbonation during the life-cycle of built and recycled concrete: Influence on their carbon footprint', *The International Journal of Life-cycle Assessment*, vol. 15, no. 6, pp. 549–56.

The Concrete Network 2016, *What makes concrete a sustainable building material?*, viewed 24.8.2017, <www.concretenetwork.com/concrete/greenbuildinginformation/what_makes.html>.

Condeixa, Karina, Haddad, Assed & Boer, Dieter 2014, 'Life-cycle impact assessment of masonry system as inner walls: A case study in Brazil', *Construction and Building Materials*, vol. 70, pp. 141–7.

Council of Australian Governments 2012, *Baseline energy consumption and greenhouse-gas emissions in commercial buildings in Australia, Part 1*, Council of Australian Governments, Australia.

Covenant of Mayors 2016, *Technical annex to the Sustainable Energy Action Plan (SEAP) template instructions document: The emission factors*, viewed 28.01.2017, <www.eumayors.eu/IMG/pdf/technical_annex_en.pdf>.

Crawley, Drury B., Hand, Jon W., Kummert, Michaël & Griffith, Brent T. 2008, 'Contrasting the capabilities of building energy performance simulation programs', *Building and Environment*, vol. 43, no. 4, pp. 661–73.

Crossin, Enda 2012, *Comparative life-cycle assessment of concrete blends*, Centre for Design, RMIT University, Melbourne, Australia.

Crowley, Thomas J. 2000, 'Causes of climate change over the past 1000 years', *Science*, vol. 289, no. 5477, pp. 270–7.

Curran, Mary Ann 2006, *Life-cycle assessment: Principles and practice*, National Risk Management Research Laboratory, Office of Research and Development, US Environmental Protection Agency, Cincinnati, OH.

Dammann, Sven & Elle, Morten 2006, 'Environmental indicators: Establishing a common language for green building', *Building Research & Information*, vol. 34, no. 4, pp. 387–404.

Damtoft, J. S., Lukasik, J., Herfort, D., Sorrentino, D. & Gartner, E. M. 2008, 'Sustainable development and climate change initiatives', *Cement and Concrete Research*, vol. 38, no. 2, pp. 115–27.

Dawood, Saad, Lord, Richard & Dawood, Nashwan 2009, 'Development of a visual whole Life-cycle energy assessment framework for built environment', in *Winter Simulation Conference, Winter Simulation Conference*, pp. 2653–63.

Day, Ken W., Aldred, James & Hudson, Barry 2013, *Concrete mix design, quality control and specification*, 4th edn, CRC Press, Boca Raton, FL.

Dean, Brian, Dulac, John, Petrichenko, Ksenia & Graham, Peter 2016, *Towards zero-emission efficient and resilient buildings: Global status report*, Global Alliance for Buildings and Construction (GABC), viewed 30.10.18, <https://orbit.dtu.dk/en/publications/towards-zero-emission-efficient-and-resilient-buildings-global-st>.

Department of Transport and Main Roads 2014, *Long distance transport and extended placement times for concrete*, Queensland Government, Australia.

Desideri, Umberto, Arcioni, Livia, Leonardi, Daniela, Cesaretti, Luca, Perugini, Perla, Agabitini, Elena & Evangelisti, Nicola 2014, 'Design of a multipurpose "zero energy consumption" building according to European Directive 2010/31/EU: Life-cycle assessment', *Energy and Buildings*, vol. 80, pp. 585–97.

De Souza, Danielle Maia, Lafontaine, Mia, Charron-Doucet, François, Chappert, Benoit, Kicak, Karine, Duarte, Fernanda & Lima, Luis 2016, 'Comparative life-cycle assessment of ceramic brick, concrete brick and cast-in-place reinforced concrete exterior walls', *Journal of Cleaner Production*, vol. 137, pp. 70–82.

Ding, Grace K. C. 2008, 'Sustainable construction: The role of environmental assessment tools', *Journal of Environmental Management*, vol. 86, no. 3, pp. 451–64.

Eastman, Chuck, Eastman, Charles M., Teicholz, Paul, Sacks, Rafael & Liston, Kathleen 2011, *BIM handbook: A guide to building information modelling for owners, managers, designers, engineers and contractors*, John Wiley & Sons, Hoboken, NJ.

Ecoinvent 2017, *Ecoinvent version 3*, Ecoinvent, viewed 30.10.2017, <www.ecoinvent.org/>.

Eichholtz, Piet, Kok, Nils & Quigley, John M. 2010, 'Doing well by doing good? Green office buildings', *The American Economic Review*, pp. 2492–509.

Emmitt, Stephen 2013, *Architectural technology: Research and practice*, John Wiley & Sons, Hoboken, NJ.

Environmental Protection Agency 2015a, *Concrete*, Environmental Protection Agency, USA, viewed, <https://www3.epa.gov/epawaste/conserve/tools/warm/pdfs/Concrete.pdf>.

Environmental Protection Agency 2015b, *Green building*, Environmental Protection Agency, USA.

European Commission 2010, *International Reference Life-cycle Data System (ILCD) handbook – Framework and requirements for life-cycle impact assessment models and indicators*, 1st edn., EUR 24586 EN., European Commission, Luxembourg, viewed, <http://eplca.jrc.ec.europa.eu/uploads/ILCD-Handbook-LCIA-Framework-Requirements-ONLINE-March-2010-ISBN-fin-v1.0-EN.pdf>.

Fan, Y. & Long, W. D. 2009, 'The carbon footprint of the HVAC system analysis and environmental evaluation', *HVAC*, vol. 39, no. 12, pp. 53–6.

Farham, Moghaddam Rad & Gholian, Mohammad Mohammad 2014, 'Leadership in energy and environmental design', *European Online Journal of Natural and Social Sciences*, vol. 3, no. 4 (s), p. 112.

Finnveden, Göran, Hauschild, Michael Z., Ekvall, Tomas, Guinée, Jeroen, Heijungs, Reinout, Hellweg, Stefanie, Koehler, Annette, Pennington, David & Suh, Sangwon 2009, 'Recent developments in life-cycle assessment', *Journal of Environmental Management*, vol. 91, no. 1, pp. 1–21.

Flower, David J. M. & Sanjayan, Jay G. 2007, 'Greenhouse-gas emissions due to concrete manufacture', *The International Journal of Life-cycle Assessment*, vol. 12, no. 5, pp. 282–8.

Flyash Australia 2010, *What is fly ash*, viewed 24.08.2017, <www.flyashaustralia.com.au/WhatIsFlyash.aspx>.

Frischknecht, Rolf, Jungbluth, Niels, Althaus, Hans-Jörg, Bauer, Christian, Doka, Gabor, Dones, Roberto, Hischier, Roland, Hellweg, S., Humbert, S. & Köllner, T. 2007, *Implementation of life-cycle impact assessment methods*, Ecoinvent Report, vol. 3.

GaBi 2017a, *Ecoinvent database*, Thinkstep, viewed 30.10.2017, <www.gabi-software.com/databases/ecoinvent/>.

GaBi 2017b, *GaBi-Software*, Thinkstep, viewed 30.10.2017, <www.gabi-software.com>.

Ga.gov.au 2016, *Maps of Australia*, Geoscience Australia, viewed 15.09.2016, <www.ga.gov.au/data-pubs/maps>.

Gaidajis, Georgios & Angelakoglou, Komninos 2011, 'Screening life-cycle assessment of an office used for academic purposes', *Journal of Cleaner Production*, vol. 19, no. 14, pp. 1639–46.

General Assembly 2015, *17 goals to transform the world*, United Nations, viewed 01.10.2017, <www.un.org/sustainabledevelopment/sustainable-development-goals/>.

Georges, Laurent, Haase, Matthias, Houlihan Wiberg, Aoife, Kristjansdottir, Torhildur & Risholt, Birgit 2015, 'Life-cycle emissions analysis of two nZEB concepts', *Building Research & Information*, vol. 43, no. 1, pp. 82–93.

Gerner, Ed & Budd, Anthony 2015, 'Australian surface temperature corrections for thermal modelling', in *Proceedings of the World Geothermal Congress 2015*.

Giurco, Damien, McLellan, Ben & Schmidt, Paul 2008, *Australian Life-cycle Initiative (AusLCI) & CSRP database: Australian data*, viewed 30.10.08, <https://opus.lib.uts.edu.au/handle/10453/20466>.

Gong, Yuanyuan & Song, Deyong 2015, 'Life-cycle building carbon emissions assessment and driving factors decomposition analysis based on IMDI – A case study of Wuhan City in China', *Sustainability*, vol. 7, no. 12, pp. 16670–86.

González, María Jesús & García Navarro, Justo 2006, 'Assessment of the decrease of CO2 emissions in the construction field through the selection of materials: Practical case study of three houses of low environmental impact', *Building and Environment*, vol. 41, no. 7, pp. 902–9.

Goodkoop, M. 1999, *The eco-indicator 99: A damage oriented method for life-cycle impact assessment*, Methodology Report.

Green Building Council of Australia 2005, *Environmental rating system for buildings*, Green Building Council of Australia, Sydney, viewed, <www.gbca.org.au/uploads/Green%20Star%20Trade%20Mark%20Rules.pdf>.

Green Building Council of Australia 2013, *The value of green star: A decade of environmental benefits*, Progress Report, vol. 194.

Green Building Council of Australia 2014a, *Green star: Design & as built, building energy consumption and greenhouse-gas emissions calculation guidelines VI*.

Green Building Council of Australia 2014b, *Materials life-cycle impacts*, Green Building Council of Australia, viewed, <www.gbca.org.au/uploads/78/34894/Materials_Life_Cycle%20Impacts_FINAL_JUNE2014.pdf>.

Green Building Council of Australia 2015a, *Green star – Design & as built*, *9770000171970*, Green Building Council of Australia, Sydney, viewed, <http://nla.gov.au/anbd.bib-an45344235>.

Green Building Council of Australia 2015b, *Greenhouse-gas emissions calculator guide*, Green Building Council of Australia, viewed, <www.gbca.org.au/uploads/195/3267/Green_Star_-_Greenhouse_Gas_Emissions_Calculator_Guide.pdf>.

Green Building Council of Australia 2015c, *Life-cycle assessment in green star*, Green Building Council of Australia, viewed 30.10.2017, <www.gbca.org.au/green-star/materials-category/life-cycle-assessment-in-green-star/>.

Green Building Council of Australia 2018, *About* GBCA, viewed 23.03.2018, <www.gbca.org.au/about/>.

GreenDelta GmbH 2017, *openLCA nexus – Your source for LCA datasets*, GreenDelta GmbH, viewed 30.10.2017, <https://nexus.openlca.org/>.

Greenhouse Gas Protocol 2011, *Product life-cycle accounting and reporting standard*, World Business Council for Sustainable Development and World Resource Institute.

Guggemos Angela, Acree & Horvath, Arpad 2005, 'Comparison of environmental effects of steel- and concrete-framed buildings', *Journal of Infrastructure Systems*, vol. 11, no. 2, pp. 93–101.

Guinée, J. 2001, 'Handbook on life cycle assessment—operational guide to the ISO standards', *The International Journal of Life Cycle Assessment*, vol. 6, no. 5, p. 255.

Hamdy, Mohamed, Nguyen, Anh-Tuan & Hensen, Jan I. M. 2016, 'A performance comparison of multi-objective optimization algorithms for solving nearly-zero-energy-building design problems', *Energy and Buildings*, vol. 121, pp. 57–71.

Hamilton, Booz Allen 2015, *Green building economic impact study*, US Green Building Council. Disponible sur le lien suivant.

Hammond, G. P. & Jones, C. I. 2008, 'Embodied energy and carbon in construction materials', *Proceedings of the Institution of Civil Engineers – Energy*, vol. 161, no. 2, pp. 87–98.

Harish, V. S. K. V. & Kumar, Arun 2014, 'Techniques used to construct an energy model for attaining energy efficiency in building: A review', in *Control, Instrumentation, Energy and Communication (CIEC), 2014 International Conference on*, IEEE, pp. 366–70.

Harish, V. S. K. V. & Kumar, Arun 2016, 'A review on modelling and simulation of building energy systems', *Renewable and Sustainable Energy Reviews*, vol. 56, pp. 1272–92.

Hassoun, M. Nadim & Al-Manaseer, Akthem 2012, *Structural concrete: Theory and design*, John Wiley & Sons, Hoboken, NJ.

Heijungs, Reinout, Guinée, Jeroen B., Huppes, Gjalt, Lankreijer, Raymond M., Udo de Haes, Helias A., Wegener Sleeswijk, Anneke, Ansems, A. M. M., Eggels, P. G., Duin, R. van & De Goede, H. P. 1992, *Environmental life-cycle assessment of products: Guide and backgrounds (part 1)*, viewed 30.10.19, <https://www.researchgate.net/publication/28645165_Environmental_Life_Cycle_Assessment_of_Products-Guide_and_Backgrounds>.

Heijungs, W. 1992, *Environmental life-cycle assessment of products-guide*, Technical Report of Centre of Environmental Science.

Henderson, Hazel 1994, 'Paths to sustainable development: The role of social indicators', *Futures*, vol. 26, no. 2, pp. 125–37.

Herrmann, Ivan T. & Moltesen, Andreas 2015, 'Does it matter which Life-cycle Assessment (LCA) tool you choose? – A comparative assessment of SimaPro and GaBi', *Journal of Cleaner Production*, vol. 86, no. Supplement C, pp. 163–9.

Hertwich, Edgar G., Mateles, Sarah F., Pease, William S. & McKone, Thomas E. 2001, 'Human toxicity potentials for life-cycle assessment and toxics release inventory risk screening', *Environmental Toxicology and Chemistry*, vol. 20, no. 4, pp. 928–39.

Hoffman, Andrew J. & Henn, Rebecca 2008, 'Overcoming the social and psychological barriers to green building', *Organization & Environment*, vol. 21, no. 4, pp. 390–419.

Holmgren, Mattias, Kabanshi, Alan & Sörqvist, Patrik 2017, 'Occupant perception of "green" buildings: Distinguishing physical and psychological factors', *Building and Environment*, vol. 114, pp. 140–7.

Hong, Jingke, Shen, Geoffrey Qiping, Feng, Yong, Lau, William Sin-tong & Mao, Chao 2015, 'Greenhouse-gas emissions during the construction phase of a building: A case study in China', *Journal of Cleaner Production*, vol. 103, pp. 249–59.

Hong, Tianzhen, Chou, S. K. & Bong, T. Y. 2000, 'Building simulation: An overview of developments and information sources', *Building and Environment*, vol. 35, no. 4, pp. 347–61.

Horvath, Arpad 2004, 'Construction materials and the environment', *Annual Review of Environment and Resources*, vol. 29, pp. 181–204.

Houghton, John T. 1996, *The science of climate change: Summary for Policymakers and Technical Summary of the Working Group I Report; Part of the Working Group I Contribution to the Second Assessment Report of the Intergovernmental Panel on Climate Change*, Cambridge University Press, Cambridge.

Houghton, John T., Meiro Filho, I. G., Callander, Bruce A., Harris, Neil, Kattenburg, Arie & Maskell, Kathy 1996, 'Climate change 1995: The science of climate change', *Climatic Change*, p. 584.

Huijbregts, Mark A. J., Steinmann, Zoran J. N., Elshout, Pieter M. F., Stam, Gea, Verones, Francesca, Vieira, Marisa, Zijp, Michiel, Hollander, Anne & van Zelm, Rosalie 2017, 'ReCiPe 2016: A harmonized life-cycle impact assessment method at midpoint and endpoint level', *The International Journal of Life-cycle Assessment*, vol. 22, no. 2, pp. 138–47.

Illankoon, I. M. Chethana S., Tam, Vivian W. Y., Le, Khoa N. & Shen, Liyin 2017a, 'Key credit criteria among international green building rating tools', *Journal of Cleaner Production*, vol. 164, pp. 209–20.

Illankoon, I. M. Chethana S., Tam, Vivian W. Y., Le, Khoa N. & Shen, Liyin 2017b, 'Key credit criteria among international green building rating tools', *Journal of Cleaner Production*, vol. 164, pp. 209–20.

Incropera, Frank P. 2011, *Fundamentals of heat and mass transfer*, 7th edn., Theodore L. Bergman et al. (eds.), John Wiley, Hoboken, NJ.

Insulation Council of Australia and New Zealand 2014, *Insulation handbook*, Insulation Council of Australia and New Zealand, Melbourne, VIC, viewed 30.06.2017, <www.qld.gov.au/environment/assets/documents/climate/climate-action-submissions/insulation-council-aus-nz-icanz.pdf>.

Intergovernmental Panel on Climate Change (IPCC) 2007, *Climate change 2007, Synbook report. Contribution of Working Groups I, II and III to the Fourth Assessment*, Report of the Intergovernmental Panel on Climate Change [Core Writing Team, Pachauri, R. K. and Reisinger, A. (eds.)], Geneva, Switzerland.

Intergovernmental Panel on Climate Change (IPCC) 2014, *Climate change 2014 synbook report summary for policymakers*, Report of the Intergovernmental Panel on Climate Change, Geneva, Switzerland.

International Building Performance Simulation Association (IBPSA) 2017, *Best directory – Building energy software tools*, viewed 30.10.17, <www.buildingenergysoftwaretools.com/>.

International Organization for Standardization 2006, *ISO 14040:2006 Environmental management – Life-cycle assessment – Principles and framework*, International Organization for Standardization Geneva, Switzerland, viewed 30.11.19, <https://www.iso.org/standard/37456.html>.

Interstate Natural Gas Association of America 2005, *Greenhouse-gas emission estimation guidelines for natural gas transmission and storage*, Volume 1 – GHG Emission Estimation Methodologies and Procedures.

Islam, Hamidul, Jollands, Margaret & Setunge, Sujeeva 2010, 'Life-cycle assessment of residential buildings: Sustainable material options in wall assemblies', in *Chemeca 2010: Engineering at the Edge*, 26–29 September 2010, Hilton Adelaide, South Australia, p. 3443.

Islam, Hamidul, Jollands, Margaret & Setunge, Sujeeva 2015a, 'Life-cycle assessment and life-cycle cost implication of residential buildings – A review', *Renewable and Sustainable Energy Reviews*, vol. 42, no. Supplement C, pp. 129–40.

Islam, Hamidul, Jollands, Margaret, Setunge, Sujeeva, Haque, Nawshad & Bhuiyan, Muhammed A. 2015b, 'Life-cycle assessment and life-cycle cost implications for roofing and floor designs in residential buildings', *Energy and Buildings*, vol. 104, pp. 250–63.

Islam, Hamidul, Zhang, Guomin, Setunge, Sujeeva & Bhuiyan, Muhammed A. 2016, 'Life-cycle assessment of shipping container home: A sustainable construction', *Energy and Buildings*, vol. 128, pp. 673–85.

Iyer-Raniga, Usha & Wong, James Pow Chew 2012, 'Evaluation of whole life-cycle assessment for heritage buildings in Australia', *Building and Environment*, vol. 47, pp. 138–49.

Jankovic, I. 2013, *Designing zero carbon buildings using dynamic simulation methods*, Taylor and Francis, Oxfordshire United Kingdom.
Japan Society of Civil Engineers 2006, *Recommendation of environmental performance verification for concrete structures*, 4810605760, Japan Society of Civil Engineers, Japan.
Jolliet, Olivier, Margni, Manuele, Charles, Raphaël, Humbert, Sébastien, Payet, Jérôme, Rebitzer, Gerald & Rosenbaum, Ralph 2003, 'IMPACT 2002+: A new life-cycle impact assessment methodology', *The international Journal of Life-cycle Assessment*, vol. 8, no. 6, p. 324.
Jørgensen, A., Herrmann, I. T. & Bjørn, A. 2013, Analysis of the link between a definition of sustainability and the life cycle methodologies. *The International Journal of Life Cycle Assessment*, vol. 18, no. 8, pp. 1440–1449.
Junnila, Seppo & Horvath, Arpad 2003, 'Life-cycle environmental effects of an office building', *Journal of Infrastructure Systems*, vol. 9, no. 4, pp. 157–66.
Kats, Gregory 2003, *Green building costs and financial benefits*, Massachusetts Technology Collaborative, Boston, MA.
Kawai, Kenji, Sugiyama, Takafumi, Kobayashi, Koichi & Sano, Susumu 2005, 'Inventory data and case studies for environmental performance evaluation of concrete structure construction', *Journal of Advanced Concrete Technology*, vol. 3, no. 3, pp. 435–56.
Khasreen, Mohamad Monkiz, Banfill, Phillip F. G. & Menzies, Gillian F. 2009, 'Life-cycle assessment and the environmental impact of buildings: A review', *Sustainability*, vol. 1, no. 3, pp. 674–701.
Khodabuccus, Rehan & Lee, Jacquetta 2016, 'A new model for designing cost effective zero carbon homes: Minimizing commercial viability issues and improving the economics for both the developer and purchaser', *Buildings*, vol. 6, no. 1, p. 6.
Kilkelly, Michael 2015, 'Five digital tools for architects to test building performance', *The Journal of the American Institute of Architects*, viewed 30.10.2017, <www.architectmagazine.com/technology/five-digital-tools-for-architects-to-test-building-performance_o>.
Kim, R. E. 2016, 'The nexus between international law and the sustainable development goals', *Review of European, Comparative & International Environmental Law*, vol. 25, no. 1, pp. 15–26.
Kline, J. & Kline, C. 2015, 'Cement and CO2: What is happening', *IEEE Transactions on Industry Applications*, vol. 51, no. 2, pp. 1289–94.
Kloepffer, Walter 2008, 'Life-cycle sustainability assessment of products', *The International Journal of Life Cycle Assessment*, vol. 13, no. 2, p. 89.
Klöpffer, Walter & Grahl, Birgit 2014, *Life-cycle assessment (LCA): A guide to best practice*, Wiley-VCH, Weinheim, Germany.
Kosmatka, Steven H., Kerkhoff, Beatrix, Panarese, William C. & Portland Cement Association 2002, *Design and control of concrete mixtures*, vol. 5420, Portland Cement Association Skokie, IL.
Kramer, Rick, van Schijndel, Jos & Schellen, Henk 2012, 'Simplified thermal and hygric building models: A literature review', *Frontiers of Architectural Research*, vol. 1, no. 4, pp. 318–25.
Kua, Harn Wei & Kamath, Susmita 2014, 'An attributional and consequential life-cycle assessment of substituting concrete with bricks', *Journal of Cleaner Production*, vol. 81, no. Supplement C, pp. 190–200.
Kua, Harn Wei & Lu, Yujie 2016, 'Environmental impacts of substituting tempered glass with polycarbonate in construction – An attributional and consequential life-cycle perspective', *Journal of Cleaner Production*, vol. 137, pp. 910–21.

Kulahcioglu, Tugba, Dang, Jiangbo & Toklu, Candemir 2012, 'A 3D analyzer for BIM-enabled life-cycle assessment of the whole process of construction', *Hvac&R Research*, vol. 18, no. 1–2, pp. 283–93.

Kumar, Mehta P. 2010, 'Sustainable cements and concrete for the climate change era – A review', in *Second International Conference on Sustainable Construction Materials and Technologies*, Citeseer.

Lawania, Krishna Kumar & Biswas, Wahidul K. 2016, 'Achieving environmentally friendly building envelope for Western Australia's housing sector: A life-cycle assessment approach', *International Journal of Sustainable Built Environment*, vol. 5, no. 2, pp. 210–24.

Lawn, Philip A. 2006, *Sustainable development indicators in ecological economics*, Edward Elgar Publishing, Cheltenham, United Kingdom.

Lawrence, K., Zeise, K., & Morgan, P. 2005, *Management options for non-road engine emissions in urban areas, consultancy report for department for environment & heritage*, Consultancy report for Department for Environment & Heritage, Canberra. Prepared by Pacific Air & Environment Pty. Ltd., Australia.

LCEE life-cycle Engineering Experts GmbH 2016, *Spun concrete columns with outstanding ecological life-cycle assessment*, Commonwealth of Australia, Germany.

Le, K. N., Tam, V. W. Y., Tran, C. N. N., Wang, J. & Goggins, B. 2018a, 'Life-cycle greenhouse-gas emission analyses for Green Star's concrete credits in Australia', *IEEE Transaction on Engineering Management*, vol. 99, pp. 1–13.

Le, K. N., Tran, C. N. N. & Tam, V. W. Y. 2018b, 'Life-cycle greenhouse-gas emissions assessment: An Australian commercial building perspective', *Journal of Cleaner Production*, vol. 199, pp. 236–47.

Lemay, Lionel 2011, *Life-cycle assessment of concrete buildings*, Concrete Sustainability Report, National Ready Mixed Concrete Association, vol. CSR04 – October 2011.

Lemay, Lionel, Lobo, Colin & Obla, Karthik 2013, 'Sustainable concrete: The role of performance-based specifications', *Structures Congress*, pp. 2693–704.

Lewis, Roland W., Nithiarasu, Perumal & Seetharamu, Kankanhalli N. 2004, *Fundamentals of the finite element method for heat and fluid flow*, John Wiley & Sons, Hoboken, NJ.

Li, Xiaodong, Zhu, Yimin & Zhang, Zhihui 2010, 'An LCA-based environmental impact assessment model for construction processes', *Building and Environment*, vol. 45, no. 3, pp. 766–75.

Life-cycle Strategies 2017, *AusLCI database*, Life-cycle Strategies, viewed 30.10.17, <www.lifecycles.com.au/auslci-database>.

Linfei, Han, Pan, Jiang & Cao, Zhikui 2011, 'The reconsideration of building Life-cycle', in *Electric Technology and Civil Engineering (ICETCE), 2011 International Conference on*, IEEE, 22–24 April 2011, pp. 4654–7.

Ma, Feng, Sha, Aimin, Yang, Panpan & Huang, Yue 2016, 'The greenhouse-gas emission from Portland cement concrete pavement construction in China', *International Journal of Environmental Research and Public Health*, vol. 13, no. 7, p. 632.

Mahbub, Rashid, Kent, Spreckelmeyer & Angrisano, Neal J. 2012, 'Green buildings, environmental awareness, and organizational image', *Journal of Corporate Real Estate*, vol. 14, no. 1, pp. 21–49.

Mamlouk, Michael S. & Zaniewski, John P. 2011, *Materials for civil and construction engineers*, Pearson Prentice Hall, Upper Saddle River, NJ.

Mangan, Suzi Dilara & Oral, Gül Koçlar 2015, 'A study on life-cycle assessment of energy retrofit strategies for residential buildings in Turkey', *Energy Procedia*, vol. 78, pp. 842–7.

May, Barrie, England, Jacqueline R., Raison, R. John & Paul, Keryn I. 2012, 'Cradle-to-gate inventory of wood production from Australian softwood plantations and native hardwood forests: Embodied energy, water use and other inputs', *Forest Ecology and Management*, vol. 264, pp. 37–50.

McKinnon, Alan & Piecyk, Maja 2010, *Measuring and managing CO2 emissions*, European Chemical Industry Council, Edinburgh, UK.

McLennan Magasanik Associates 2010, *Climate change and the resource recovery and waste sectors*, A report prepared for the Department of the Environment, Water, Heritage and the Arts, Melbourne, Australia.

Menke, Dean M., Davis, Gary A. & Vigon, Bruce W. 1996, *Evaluation of life-cycle assessment tools*, Environment Canada Gatineau, Quebec, Canada.

Mindess, Sidney, Young, J. Francis & Darwin, David 2003, *Concrete*, Prentice Hall, Upper Saddle River, NJ.

Mitalas, G. P. & Stephenson, Donald George 1967, *Room thermal response factors*, viewed 20.12.19, <http://web.mit.edu/parmstr/Public/NRCan/nrcc23302.pdf>.

Molenbroek, E., Smith, M., Surmeli, N., Schimschar, S., Waide, P., Tait, J. & McAllister, C. 2015, *Savings and benefits of global regulations for energy efficient products: A "cost of non-world" study*, European Commission, Brussels.

Morel, J. C., Mesbah, A., Oggero, M. & Walker, P. 2001, 'Building houses with local materials: Means to drastically reduce the environmental impact of construction', *Building and Environment*, vol. 36, no. 10, pp. 1119–26.

Morrissey, J. & Horne, R. E. 2011, 'Life-cycle cost implications of energy efficiency measures in new residential buildings', *Energy and Buildings*, vol. 43, no. 4, pp. 915–24.

Müller, Nicolas & Harnisch, Jochen 2008, 'A blueprint for a climate friendly cement industry', *Report for the WWF – Lafarge Conservation Partnership*, WWF, Gland, Switzerland.

Mustafaraj, G., Lowry, G. & Chen, J. 2011, 'Prediction of room temperature and relative humidity by autoregressive linear and nonlinear neural network models for an open office', *Energy and Buildings*, vol. 43, no. 6, pp. 1452–60.

Myhre, G., Shindell, D., Bréon, F. M., Collins, W., Fuglestvedt, J., Huang, J. & Zhang, H. 2013, *Anthropogenic and natural radiative forcing. Climate Change 2013: The Physical Science Basis. Contribution of Working Group I to the Fifth Assessment Report of the Intergovernmental Panel on Climate Change*, 659–740, Cambridge University Press, Cambridge.

National Laboratory of the U.S. Department of Energy 2017, *OpenStudio*, NREL, viewed 30.09.2017, <www.openstudio.net/>.

National Renewable Energy laboratory 2012, *U.S. life-cycle inventory database*, National Renewable Energy Laboratory, viewed 30.10.2017, <www.lcacommons.gov/nrel/search>.

Neil Adger, W., Arnell, Nigel W. & Tompkins, Emma L. 2005, 'Successful adaptation to climate change across scales', *Global Environmental Change*, vol. 15, no. 2, pp. 77–86.

Neville, Adam M. 1997, 'Aggregate bond and modulus of elasticity of concrete', *ACI Materials Journal*, vol. 94, pp. 71–8.

Neville, Adam M. & Brooks, J. J. 2010, *Concrete technology*, 2nd edn, Pearson United Kingdom, Edinburgh, England.

Nicol, Sam & Chadès, Iadine 2017, 'A preliminary approach to quantifying the overall environmental risks posed by development projects during environmental impact assessment', *PLoS ONE*, vol. 12, no. 7, p. e0180982.

Nicola, Dempsey, Glen, Bramley, Sinéad, Power & Caroline, Brown 2011, 'The social dimension of sustainable development: Defining urban social sustainability', *Sustainable Development*, vol. 19, no. 5, pp. 289–300.

Nielsen, Claus Vestergaard 2008, 'Carbon footprint of concrete buildings seen in the life-cycle perspective', in *NRMCA 2008 Concrete Technology Forum*.

Olivier, Jos G. J., Janssens-Maenhout, Greet, Muntean, Marilena & Peters, J. A. H. W. 2016, *Trends in global CO2 emissions; 2016 report*, Environmental Assessment Agency, PBL Netherlands, The Hague.

Olubunmi, Olanipekun Ayokunle, Xia, Paul Bo & Skitmore, Martin 2016, 'Green building incentives: A review', *Renewable and Sustainable Energy Reviews*, vol. 59, pp. 1611–21.

O'Malley, Christopher, Piroozfar, Poorang A. E., Farr, Eric R. P. & Gates, Jonathan 2014, 'Evaluating the efficacy of BREEAM Code for Sustainable Homes (CSH): A cross-sectional study', *Energy Procedia*, vol. 62, pp. 210–19.

Omer, Abdeen Mustafa 2008, 'Energy, environment and sustainable development', *Renewable and Sustainable Energy Reviews*, vol. 12, no. 9, pp. 2265–300.

Organisation for Economic Co-operation and Development (OECD) 2017, *Greenhouse-gas emissions 1994–2015*, viewed 12.09.2017, <https://stats.oecd.org/Index.aspx?DataSetCode=AIR_GHG>.

Ortiz-Rodríguez, Oscar, Castells, Francesc & Sonnemann, Guido 2009, 'Sustainability in the construction industry: A review of recent developments based on LCA', *Construction and Building Materials*, vol. 23, no. 1, pp. 28–39.

Ortiz-Rodríguez, Oscar, Castells, Francesc & Sonnemann, Guido 2010, 'Life-cycle assessment of two dwellings: One in Spain, a developed country, and one in Colombia, a country under development', *Science of the Total Environment*, vol. 408, no. 12, pp. 2435–43.

Owsianiak, Mikołaj, Laurent, Alexis, Bjørn, Anders & Hauschild, Michael Z. 2014, 'IMPACT 2002+, ReCiPe 2008 and ILCD's recommended practice for characterization modelling in life-cycle impact assessment: A case study-based comparison', *The International Journal of Life-cycle Assessment*, vol. 19, no. 5, pp. 1007–21.

Ozisik, Necati 1994, *Finite difference methods in heat transfer*, CRC Press, Boca Raton, FL.

Papadopoulos, A. M. 2005, 'State of the art in thermal insulation materials and aims for future developments', *Energy and Buildings*, vol. 37, no. 1, pp. 77–86.

Percival, Robert V., Schroeder, Christopher H., Miller, Alan S. & Leape, James P. 2017, *Environmental regulation: Law, science, and policy*, Wolters Kluwer Law & Business, New York, NY.

Perera, D. W. U., Winkler, D. & Skeie, N.-O. 2016, 'Multi-floor building heating models in MATLAB and Modelica environments', *Applied Energy*, vol. 171, pp. 46–57.

Puettmann, Maureen E. & Wilson, James B. 2007, 'Life-cycle analysis of wood products: Cradle-to-gate LCI of residential wood building materials', *Wood and Fiber Science*, vol. 37, pp. 18–29.

Ramage, Michael H., Burridge, Henry, Busse-Wicher, Marta, Fereday, George, Reynolds, Thomas, Shah, Darshil U., Wu, Guanglu, Yu, Li, Fleming, Patrick, Densley-Tingley, Danielle, Allwood, Julian, Dupree, Paul, Linden, P. F. & Scherman, Oren 2017, 'The wood from the trees: The use of timber in construction', *Renewable and Sustainable Energy Reviews*, vol. 68, pp. 333–59.

Ramesh, T., Prakash, Ravi & Shukla, K. K. 2010, 'Life-cycle energy analysis of buildings: An overview', *Energy and Buildings*, vol. 42, no. 10, pp. 1592–600.

Rehan, R. & Nehdi, M. 2005, 'Carbon dioxide emissions and climate change: Policy implications for the cement industry', *Environmental Science & Policy*, vol. 8, no. 2, pp. 105–14.

Renouf, M. A., Grant, T., Sevenster, M., Logie, J., Ridoutt, B., Ximenes, F., Bengtsson, J., Cowie, A. & Lane, J. 2015, *Best practice guide for Life-cycle Impact Assessment (LCIA) in Australia*, Australian Life-cycle Assessment Society, Melbourne, viewed 20.10.19,

<http://www. auslci. com. au/Documents/Best_Practice_Guide_V2_Draft_for_Consultation. pdf (2015)>.

Robertson, Adam B., Lam, Frank C. F. & Cole, Raymond J. 2012, 'A comparative cradle-to-gate Life-cycle assessment of mid-rise office building construction alternatives: Laminated timber or reinforced concrete', *Buildings*, vol. 2, no. 3, pp. 245–70.

Robichaud, Lauren Bradley & Anantatmula, Vittal S. 2011, 'Greening project management practices for sustainable construction', *Journal of Management in Engineering*, vol. 27, no. 1, pp. 48–57.

Rodríguez, C. & Ciroth, A. 2016, *Adapting LCA software to LCI databases and vice versa*, Ökobilanzwerkstatt, Pforzheim.

Sabnis, Gajanan M. 2015, *Green building with concrete: Sustainable design and construction*, CRC Press, Boca Raton, FL.

Sachs, Jeffrey D. 2012, 'From millennium development goals to sustainable development goals', *The Lancet*, vol. 379, no. 9832, pp. 2206–11.

Salcido, Juan C., Raheem, Adeeba Abdul & Ravi, Srinivasan 2016, 'Comparison of embodied energy and environmental impact of alternative materials used in reticulated dome construction', *Building and Environment*, vol. 96, pp. 22–34.

Seo, Seongwon, Kim, Junbeum, Yum, Kwok-Keung & McGregor, James 2015, 'Embodied carbon of building products during their supply chains: Case study of aluminium window in Australia', *Resources, Conservation and Recycling*, vol. 105, pp. 160–6.

Seppälä, Jyri, Posch, Maximilian, Johansson, Matti & Hettelingh, Jean-Paul 2006, 'Country-dependent characterisation factors for acidification and terrestrial eutrophication based on accumulated exceedance as an impact category indicator (14 pp)', *The International Journal of Life-cycle Assessment*, vol. 11, no. 6, pp. 403–16.

Shahiduzzaman, M. D. & Layton, Allan 2015, 'Decomposition analysis to examine Australia's 2030 GHGs emissions target: How hard will it be to achieve?', *Economic Analysis and Policy*, vol. 48, no. Supplement C, pp. 25–34.

Sharma, Aashish, Saxena, Abhishek, Sethi, Muneesh & Shree, Venu 2011, 'Life-cycle assessment of buildings: A review', *Renewable and Sustainable Energy Reviews*, vol. 15, no. 1, pp. 871–5.

The Silica Fume Association 2016, *What is silica fume?*, viewed 24.08.2017, <www.silicafume.org/general-silicafume.html>.

Sim, Jaehun, Sim, Jehean & Park, Changbae 2016, 'The air emission assessment of a South Korean apartment building's life-cycle, along with environmental impact', *Building and Environment*, vol. 95, pp. 104–15.

SimaPro 2017a, *Ecoinvent LCI database*, SimaPro, viewed 30.10.2017, <https://simapro.com/databases/ecoinvent/>.

SimaPro 2017b, *SimaPro-Software*, SimaPro, viewed 30.10.2017, <www.simapro.de>.

Singh, Amanjeet, Berghorn, George, Joshi, Satish & Syal, Matt 2010, 'Review of life-cycle assessment applications in building construction', *Journal of Architectural Engineering*, vol. 17, no. 1, pp. 15–23.

Sinha, Rajib, Lennartsson, Maria & Frostell, Björn 2016, 'Environmental footprint assessment of building structures: A comparative study', *Building and Environment*, vol. 104, pp. 162–71.

Soubbotina, Tatyana P. 2004, *Beyond economic growth: An introduction to sustainable development*, The World Bank, Washington, D.C.

Speck, Ricky, Selke, Susan, Auras, Rafael & Fitzsimmons, James 2016, 'Life-cycle assessment software: Selection can impact results', *Journal of Industrial Ecology*, vol. 20, no. 1, pp. 18–28.

Standards Association of Australia 2007, *Specification and supply of concrete*, 3rd ed., Standards Australia (Standards Association of Australia), North Sydney, NSW, Australia.

Stephenson, Donald George & Mitalas, G. P. 1967, 'Cooling load calculations by thermal response factor method', *ASHRAE Transactions (United States)*, vol. 73, no. 1, pp. 1–7.

Stephenson, Donald George & Mitalas, G. P. 1971, 'Calculation of heat conduction transfer functions for multi-layers slabs', *Air Cond. Engrs. Trans (United States)*, vol. 77, no. 2, 117–126.

Taborianski, Vanessa Montoro & Prado, Racine T. A. 2004, 'Comparative evaluation of the contribution of residential water heating systems to the variation of greenhouse-gases stock in the atmosphere', *Building and Environment*, vol. 39, no. 6, pp. 645–52.

Tam, V. W., Le, K. N., Evangelista, A. C. J., Butera, A., Tran, C. N. & Teara, A. 2019, 'Effect of fly ash and slag on concrete: Properties and emission analyses', *Frontiers of Engineering Management*, vol. 6, no. 3, pp. 395–405.

Tam, V. W. Y., Le, K. N., Senaratne, S., Shen, I. Y., Perica, J. & Illankoon, I. C. S. 2017a, 'Life-cycle cost analysis of green-building implementation using timber applications', *Journal of Cleaner Production*, vol. 147, no. Supplement C, pp. 458–69.

Tam, V. W. Y., Le, K. N. & Tran, C. N. N. 2017b, 'A review on international ecological legislations on life-cycle energy consumption, greenhouse-gas emissions and costing assessment', in *Proceedings of 22nd International Conference on Advancement of Construction Management and Real Estate*, 'CRIOCM 2017', CRIOCM 2017 Organising Committee, Melbourne, 20th–23rd November 2017.

Tam, V. W. Y., Le, K. N. & Tran, C. N. N. 2017d, 'An exploratory model on greenhouse-gas emissions and costing assessment in building operation stage', in *Conference on Innovative Production and Construction 2017*, Perth, 30th November–1st December 2017.

Tam, V. W. Y., Le, K. N. & Tran, C. N. N. 2018b, 'Optimizing Life-cycle carbon emissions for achieving concrete credits in Australia', in K. W. Chau, I. Y. S. Chan, W. Lu & C. Webster (eds), *Proceedings of the 21st international symposium on advancement of construction management and real estate*, Springer Singapore, Singapore, pp. 1077–87.

Tam, V. W. Y., Le, K. N., Tran, C. N. N. & Wang, J. Y. 2018a, 'A review on contemporary computational programs for building's life-cycle energy consumption and greenhouse-gas emissions assessment', *Journal of Cleaner Production;*, vol. 172, pp. 4220–30.

Tam, V. W. Y., Le, K. N., Tran, C. N. N., Wang, X. & Wang, J. Y. 2017c, 'A review of international green building designs', *International Journal of Construction Project Management*, vol. 9, no. 1, pp. 3–18.

Tam, V. W. Y. & Tam, C. M. 2008, *Re-use of construction and demolition waste in housing developments*, Nova Science Publishers, New York, United States.

Taylor, J. & Van Langenberg, K. 2003, *Review of the environmental impact of wood compared with alternative products used in the production of furniture*, Forest & Wood Products Research & Development Corporation, World Trade Centre, Victoria, Australia.

Techato, Kua-anan, Watts, Daniel J. & Chaiprapat, Sumate 2009, 'Life-cycle analysis of retrofitting with high energy efficiency air-conditioner and fluorescent lamp in existing buildings', *Energy Policy*, vol. 37, no. 1, pp. 318–25.

Thomson, Giles, Matan, Annie & Newman, Peter 2013, 'A review of international low carbon precincts to identify pathways for mainstreaming Sustainable urbanism in Australia', in *SOAC Conference 2013 Proceedings*, State of Australian Cities Research Network.

Trimble Inc. 2017, *Buy SketchUp Pro 2017*, Trimble Inc., viewed 30.08.2017, <www.sketchup.com/buy/sketchup-pro>.

United Nations 2007, *Indicators of sustainable development: Guidelines and methodologies*, United Nations New York.

United States Environment Protection Agency 2014, *Green building*, viewed 29.09.2015, <http://archive.epa.gov/greenbuilding/web/html/about.html>.

United States Green Building Council 2007, *Making the business case for high performance green buildings*, 1800 Massachusetts Ave, NW Suite 300 Washington, DC 20036.

United States Green Building Council 2015, *LEED*, United States Green Building Council, viewed 01.08.2015, <www.usgbc.org/>.

U.S. Energy Information Administration 2016, *International energy outlook 2016: With projections to 2040*, U.S. Energy Information Administration (EIA), USA.

Vahidi, Ehsan & Zhao, Fu 2017, 'Environmental life-cycle assessment on the separation of rare earth oxides through solvent extraction', *Journal of Environmental Management*, vol. 203, pp. 255–63.

Vandecasteele, Ine, Marí Rivero, Inés, Sala, Serenella, Baranzelli, Claudia, Barranco, Ricardo, Batelaan, Okke & Lavalle, Carlo 2015, 'Impact of shale gas development on water resources: A case study in Northern Poland', *Environmental Management*, vol. 55, no. 6, pp. 1285–99.

Van Vuuren, Detlef P., Stehfest, Elke, Gernaat, David E. H. J., Doelman, Jonathan C., Van den Berg, Maarten, Harmsen, Mathijs, de Boer, Harmen Sytze, Bouwman, Lex F., Daioglou, Vassilis & Edelenbosch, Oreane Y. 2017, 'Energy, land-use and greenhouse-gas emissions trajectories under a green growth paradigm', *Global Environmental Change*, vol. 42, pp. 237–50.

Van Zelm, Rosalie, Huijbregts, Mark A. J., den Hollander, Henri A., Van Jaarsveld, Hans A., Sauter, Ferd J., Struijs, Jaap, Van Wijnen, Harm J. & Van de Meent, Dik 2008, 'European characterization factors for human health damage of PM 10 and ozone in life-cycle impact assessment', *Atmospheric Environment*, vol. 42, no. 3, pp. 441–53.

Victoria WorkSafe 2008, *Workplace amenities and work environment*, Victoria Government: Minister for Finance, Work Cover and the Transport Accident Commission.

Vigon, Bruce 1993, *Guidelines for life-cycle assessment: A code of practice*, Society of Environmental Toxicology and Chemistry (SETAC), Pensacola, Florida, USA.

Villoria-Saez, Tam, Merino, del Río, Arrebola & Wang, X. 2016, 'Effectiveness of greenhouse-gas emission trading schemes implementation: A review on legislations', *Journal of Cleaner Production*, vol. 127, pp. 49–58.

Virag, Zdravko, Živić, Marija & Krizmanić, Severino 2011, 'Cooling of a sphere by natural convection – The applicability of the lumped capacitance method', *International Journal of Heat and Mass Transfer*, vol. 54, no. 11, pp. 2303–9.

Wang, Z. Z., Fan, I. C. & Mark, H. 2011, 'Life-cycle assessment of CO2 emissions of buildings', in *2011 International Conference on Remote Sensing, Environment and Transportation Engineering*, 24–26 June 2011, pp. 438–41.

Wang, Zhaohua, Liu, Wei & Yin, Jianhua 2015, 'Driving forces of indirect carbon emissions from household consumption in China: An input–output decomposition analysis', *Natural Hazards*, vol. 75, no. 2, pp. 257–72.

Wang, Zhaohua & Wang, Chen 2015, 'How carbon offsetting scheme impacts the duopoly output in production and abatement: Analysis in the context of carbon cap-and-trade', *Journal of Cleaner Production*, vol. 103, pp. 715–23.

Weißenberger, Markus, Jensch, Werner & Lang, Werner 2014, 'The convergence of life-cycle assessment and nearly zero-energy buildings: The case of Germany', *Energy and Buildings*, vol. 76, pp. 551–7.

Weidema, Bo, Wenzel, Henrik, Petersen, Claus & Hansen, Klaus 2004, 'The product, functional unit and reference flows in LCA', *Environmental News*, vol. 70, pp. 1–46.

Willis, Katie and Gangell, Simone 2012, 'Profiling heavy vehicle speeding', *Trends & Issues in Crime and Criminal Justice*, no. 446.

Winston, Nessa 2010, 'Regeneration for sustainable communities? Barriers to implementing sustainable housing in urban areas', *Sustainable Development*, vol. 18, no. 6, pp. 319–30.

Wong, Peter S. P., Lacarruba, Nicholas & Bray, Adam 2012, 'Can a carbon tax push the Australian construction sector toward self-regulation? Lessons learned from European union experiences', *Journal of Legal Affairs and Dispute Resolution in Engineering and Construction*, vol. 5, no. 4, pp. 163–7.

Wong, Sheau-Chyng & Abe, Naoya 2014, 'Stakeholders' perspectives of a building environmental assessment method: The case of CASBEE', *Building and Environment*, vol. 82, pp. 502–16.

World Meteorological Organization (WMO) 2017, *Greenhouse-gas concentrations surge to new record*, World Meteorological Organization (WMO), viewed 30.10.2017, <https://public.wmo.int/en/media/press-release/greenhouse-gas-concentrations-surge-new-record>.

World Wildlife Fund 2015, *Green building design*, viewed 28.09.2015, <www.wwf.org.au/get_involved/change_the_way_you_live/green_building_design/>.

Wu, Peng, Xia, Bo & Zhao, Xianbo 2014, 'The importance of use and end-of-life phases to the Life-cycle greenhouse-gas (GHG) emissions of concrete – A review', *Renewable and Sustainable Energy Reviews*, vol. 37, pp. 360–9.

Ximenes, Fabiano A. & Grant, Tim 2013, 'Quantifying the greenhouse benefits of the use of wood products in two popular house designs in Sydney, Australia', *The International Journal of Life-cycle Assessment*, vol. 18, no. 4, pp. 891–908.

Yuan, Yan & Jin, Zhonghua 2015, 'Life-cycle assessment of building energy in big-data era: Theory and framework', in *Network and Information Systems for Computers (ICNISC), 2015 International Conference on*, IEEE, pp. 601–5.

Zabalza Bribián, Ignacio, Aranda Usón, Alfonso & Scarpellini, Sabina 2009, 'Life-cycle assessment in buildings: State-of-the-art and simplified LCA methodology as a complement for building certification', *Building and Environment*, vol. 44, no. 12, pp. 2510–20.

Zabalza Bribián, Ignacio, Valero Capilla, Antonio & Aranda Usón, Alfonso 2011, 'Life-cycle assessment of building materials: Comparative analysis of energy and environmental impacts and evaluation of the eco-efficiency improvement potential', *Building and Environment*, vol. 46, no. 5, pp. 1133–40.

Zelm, R. 2009, 'ReCiPe 2008, a life-cycle impact assessment method which comprises harmonised category indicators at the midpoint and the endpoint level. Report I: Characterisation; 6 January 2009', *Resources, Conservation & Recycling*, vol. 49, no. 1, p. 3248.

Zhang, Chunxia, Zhang, Peipei & Huang, Youliang 2010, 'Building energy carbon emission factor selection method', *Construction Economics*, vol. 10, pp. 106–9.

Zhang, Xiaoling, Shen, Liyin & Wu, Yuzhe 2011, 'Green strategy for gaining competitive advantage in housing development: A China study', *Journal of Cleaner Production*, vol. 19, no. 2, pp. 157–67.

Zhang, Xiaocun & Wang, Fenglai 2015, 'Life-cycle assessment and control measures for carbon emissions of typical buildings in China', *Building and Environment*, vol. 86, pp. 89–97.

Zhu, Jianjun, Chew, David A. S., Lv, Sainan & Wu, Weiwei 2013, 'Optimization method for building envelope design to minimize carbon emissions of building operational energy consumption using Orthogonal Experimental Design (OED)', *Habitat International*, vol. 37, pp. 148–54.

Židonienė, Sigita & Kruopienė, Jolita 2015, 'Life-cycle assessment in environmental impact assessments of industrial projects: Towards the improvement', *Journal of Cleaner Production*, vol. 106, pp. 533–40.

Zimmermann, M., Althaus, H. J. & Haas, A. 2005, 'Benchmarks for sustainable construction: A contribution to develop a standard', *Energy and Buildings*, vol. 37, no. 11, pp. 1147–57.

Index

Printed in the United States
by Baker & Taylor Publisher Services